Lecture Notes
in Control and Information Sciences   356

Editors: M. Thoma, M. Morari

Andrei Karatkevich

Dynamic Analysis of Petri Net-Based Discrete Systems

 Springer

Author

Dr. Andrei Karatkevich
Institute of Computer Engineering and Electronics
Faculty of Electrical Engineering,
Computer Science and Telecommunications
University of Zielona Góra
Ul. Pogfórna 50
65-246 Zielona Góra
Poland
Email: A.Karatkevich@IIE.UZ.ZGORA.PL

Library of Congress Control Number: 2007923719

ISSN print edition: 0170-8643
ISSN electronic edition: 1610-7411
ISBN-10   3-540-71464-2 Springer Berlin Heidelberg New York
ISBN-13   978-3-540-71464-4 Springer Berlin Heidelberg New York

Springer is a part of Springer Science+Business Media
springer.com
© Springer-Verlag Berlin Heidelberg 2007

Typesetting: by the authors and SPS using a Springer LaTeX macro package

SPIN: 11903901   89/SPS   5 4 3 2 1 0

Preface

Design of modern digital hardware systems and of complex software systems is almost always connected with parallelism. For example, execution of an object-oriented program can be considered as parallel functioning of the co-operating objects; all modern operating systems are multitasking, and the software tends to be multithread; many complex calculation tasks are solved in distributed way. But designers of the control systems probably have to face parallelism in more evident and direct way. Controllers rarely deal with just one controlled object. Usually a system of several objects is to be controlled, and then the control algorithm naturally turns to be parallel.

So, classical and very deeply investigated model of discrete device, Finite State Machine, is not expressive enough for the design of control devices and systems. Theoretically in most of cases behavior of a controller can be described by an FSM, but usually it is not convenient; such FSM description would be much more complex, than a parallel specification (even as a network of several communicating FSMs).

The engineers and researchers became aware of practical necessity of developing of parallel discrete models about forty years ago. There were (and are) two main approaches to such models. One is a direct development of the FSM, being enhanced by parallelism and hierarchy. Another one is based on the parallelism "from the very beginning". The most famous and popular model of this second kind is Petri nets. A big family of more or less detailed behavioral specifications of parallel systems is based on this formalism.

Design and, especially, analysis and verification of systems, which behavior is specified by the parallel models, is a remarkably more complex task, than design and analysis of strictly sequential systems.

In this book, we are concerned about the formal analysis and verification of the parallel systems, specified by the Petri nets and the extended Petri net models. Besides, some results presented here are related to the FSM networks (but, again, we model them by means of Petri nets) and to the sequent automata (a kind of parallel descriptions other than Petri nets). To formulate some general affirmations, we use a general model of a parallel discrete system, which covers all specific models studied in the book.

We have focused on the approach of reduced exploration of state spaces. This approach is selected here as the basic one, because the state exploration provides the most detailed information about system behavior among other analysis approaches, and, on the other hand, such exploration does not have to be full to decide many important properties of the systems (especially with restricted structure).

The analysis methods of such kind are thoroughly developed; our work was inspired by the results of many authors, first of all of A. Valmari and P. Godefroid. For inspiration of another kind (the interesting parallel models and the methodology of research) we are grateful to A. Zakrevskij. The original results presented in this book are mostly connected to generalization of the known methods or, vice versa, to applying them to specific subclasses of parallel systems, which sometimes allows to obtain more information than in general case.

We intended this book to be useful to CAD researches and designers of parallel control systems. Content of the book is mostly theoretical, but it was written bearing in the mind possible practical applications. It may also be useful for the students of electrical engineering and computer science.

March 2007   Andrei Karatkevich
                                                                        Zielona Góra

Symbols

Main symbols

A	FSM or parallel automaton
C	incidence matrix of a Petri net
D	siphon
e	event
E	set of edges of a graph
F	set of arcs of a Petri net
G	graph
I	set of input (external) events
k	elementary conjunction; implicant of a Boolean function
L	cycle in a graph
M	marking or global state
M_d	deadlock
M_0	initial marking or initial global state
mp	macroplace
N	FSM network
O	set of output events
p	place or local state
$\mathbf{p}_i$	Boolean variable corresponding to place or local state p_i
P	set of places or set of local states
P^{in}	set of input places
P^{out}	set of output places
Q	path in a graph
Qp_i	code of local state p_i (an elementary conjunction)
R	partial order relation
s	sequent
S	sequent automaton
t	transition
T	set of transitions
T_P	persistent set
T_S	stubborn set

u	elementary disjunction
V	set of nodes of a graph
w	weight (of an arc)
X	set of input variables
Y	set of output variables
Z	set of internal variables or events
Γ	set of events
Δ	step of concurrent simulation
ϵ_i	Boolean variable corresponding to event e_i
μ	initial label of a parallel automaton transition
ν	terminal label of a parallel automaton transition
ρ	priority relation
σ	firing sequence
Σ	Petri net
φ	left part of a sequent
ψ	right part of a sequent

Main operators and functions

$enabled(M)$	set of transitions enabled in M
$M(p)$	number of tokens in place p or activity of local state p at M
$M(P)$	sum of tokens in places belonging to P at M
$[M\rangle$	set of markings or global states, reachable from M
$P(p)$	set of local states parallel to p
$V(G)$	set of nodes of graph G
$^\bullet x$	set of predecessors of node (place or transition) x of a Petri net
$x^\bullet$	set of successors of node (place or transition) x of a Petri net
$[x]$	cluster of Petri net containing node x
$|\sigma|$	length of firing sequence σ

Main abbreviations

BDD	binary decision diagram
CAD	computer aided design
CNF	conjunctive normal form
DNF	disjunctive normal form
EFC	extended free choice
FSM	finite state machine
HPN	hierarchical Petri net
IPN	interpreted Petri net
JPVM	Java Parallel Virtual Machine
LS	live and safe
OPN	operational Petri net
OPT	"optimal simulation"
PN	Petri net

PNSF	Petri Net Specification Format
PSS	"parallel selective search"
PDG	program dependency graph
RRG	reduced reachability graph
SCC	strongly connected component
SFC	Sequential Function Chart
SM	state machine
TC	terminal component
UML	Unified Modelling Language
XML	Extensible Markup Language

Contents

1. Introduction

Design of modern discrete devices and systems often deals with parallel processes and structures. For that reason practically all modern hardware design languages and formalisms used for system specification (such as VHDL, Verilog) allow describing concurrency. Design of complex, VLSI-based electronic devices is possible only with the help of CAD systems, so the design and verification methods have to be (and mostly are) formalized. Formalization and automatization of system design requires developing of formal models for parallel discrete systems and low-level description languages based on these models.

Specifications of devices and systems described in VHDL, Verilog or other popular languages of logical control, as LD, IL or ST [159], are very difficult for formal verification, because it is practically impossible to create adequate and at the same time simple formal models for such specifications (if these languages are used without restrictions). The problem can be solved by using of restricted specifications based on models which are easy to analyze and have enough expressive power.

There are two main directions of developing such models, each having its good and bad aspects. Both of them are, in a sense, extensions of the finite state machines (FSM) - the basic model of sequential discrete devices, which is, of course, in its "pure" version not convenient for practical needs of specifying of complex systems.

One direction is the composition of FSMs in various ways. The simplest implementation of this approach is the FSM network - a system of communicating automata [23, 147]. Studies on the automata networks have started in 1960-s, but rapid development of the methods of behavior specification by means of such networks began in 1980-s. Adding hierarchy to FSM networks leads to obtaining the model known as HCFSM (Hierarchical Concurrent Finite State Machines) [71]. One of the most popular and well-adapted to HCFSM languages has been developed within a frame of the universal specification language UML (Unified Modelling Language [209]), describing hierarchical objects and dependencies between them. We talk about the Statechars, invented by D. Harel [56, 83]. There exist several other models and languages based on automata networks such as

A. Karatkevich: Dynamic Analysis of Petri Net-Based Discrete Sys., LNCIS 356, pp. 1–8, 2007.
springerlink.com   © Springer-Verlag Berlin Heidelberg 2007

SMV [177], Promela [89, 90], CFSM [18], Requirements State Machine Language and visualState [200]. Probably the best implementation of this approach is the Ptolemy project, developed in Berkeley university [52, 157].

Another direction is the Petri nets and Petri net-based models and languages. A "pure" Petri net can describe a structure of parallel algorithm in convenient way, but it cannot describe interaction with the outer world (controlled objects). In order to develop Petri net models useful for discrete system design, the nets have to be enhanced at least by input and output signals. Often also such elements, as internal signals, time dependencies, operations on integers and other non-binary data are used. Probably the first successful attempt to create a Petri net-based language for control specification was the model known as GRAFCET. Its first version was developed in France by the working group called "Logical Systems" from AFCET (Association Française de Cybernétique Economique et Technique) in the 1970s [51, 205]. In 1988 it was adopted by the International Electrotechnical Commission as an international standard under the name of "Sequential Function Chart" (SFC)[159]. Translators have been developed to implement GRAFCET on programmable controllers. In the late 1970s and the early 1980s intensive researches in similar direction have started in the USSR (Institute of Engineering Cybernetics, Minsk [236, 238, 239, 240]; Institute of Control Sciences, Moscow [232, 233]) and in Poland (Technical University of Zielona Góra [3, 224]); parallel automata models [4, 240] and PRALU language [245, 249] arose, the methods of implementation and verification of such descriptions were designed[1]. Later various kinds of colored, interpreted, object and hybrid Petri nets and similar models have been developed, studied and applied (see e.g. [12, 34, 67, 94, 95, 100, 137, 143, 195, 196]). Most of these models allow hierarchical description, which is necessary for modern system design. In this case a net place or (more rarely) a net transition may be considered at lower level as a net.

These two approaches are equivalent in their expressiveness, each of them has its ardent supporters, the Petri net models and FSM-based models can be transformed into each other [148], and the question, which of them is "better" for system design, is still discussed. Both of them are used in CAD systems (see, for example, [56, 253]; however, Petri net models, being popular among the researchers, are definitively less popular among industrial CAD designers), and there is practical need to develop analysis methods for the models of both kinds.

A control system can be implemented using one or several microcontrollers, FPGA devices, specialized or general-purpose computers and so on. At the level of control algorithm specification and its verification there is no difference, which way of implementation will be used at further steps of design. So, in this book analysis and verification tasks are considered independently of the implementation details.

---

[1] These researches were preceded by studying the *sequent descriptions* [4, 80, 234, 237, 250], probably inspired by the theory of logical inference introduced by Gentzen [204]. Now sequent descriptions (sequent automata) are used as one of the intermediate specifications during implementation of parallel automata [5, 243, 248, 249].

1.1 Analysis of Parallel Discrete Systems

Methods of formal verification are a necessary part of any methodology of computer-aided design of hardware or software systems - and, in most cases, the verification is a bottleneck of the designs. Verification of parallel safety-critical systems differs from the verification of sequential designs, because there are some additional important properties, guaranteed or easy to check in case of sequential systems. The main conditions of a "good" parallel system are the next: [200, 245, 249]:

- lack of redundancy (lack of unreachable local states[2] and operations which are never executed);
- deadlock-freeness (in some cases the specified deadlocks must be reachable in a system; generally, detection of global and local deadlocks - situations, in which some or all parts of the system cannot react on input events because of mutual blocking - is one of the main analysis tasks) and a wider property, liveness (which implies lack of redundancy); often (for the cyclic systems) the condition of reversibility is added;
- safeness - no operation can be re-initialized during its execution;
- determinism - for parallel systems it additionally means, that parallel branches never destroy conditions and results of each other.

Of course, a designer may choose and formalize some specific conditions for specific designs. Checking of most conditions can be reduced to solving of reachability problem (reachability of a specified state or one of the states belonging to a specified class). Reachability is usually not a property difficult to check for a sequential system such as FSM, but for parallel systems the situation is different.

Analysis and formal verification of the parallel systems is a much more complex task than verification of a single FSM. The main problem is caused by the fact, that the parallel systems may have huge number of reachable states (it may depend exponentially on the of system size; for example, number of states of an FSM, equivalent to a parallel automaton with n states, may be maximally $3^{n/3}/n$ [249]. A parallel system may even have the infinite state space, such as an unbounded Petri net). That's why analysis by reducing of a parallel system to sequential one or by generating its state space in explicit form is practically impossible even for relatively simple systems (a parallel automaton with several dozens of local states may have milliards of global spaces). Model checking, a popular technology of formal verification based on state space analysis (using some tricks to prune the state space and, in most cases, representing state space in compact form like BDD), is practically used to verify properties of state spaces of size as large as about 1000 states maximum (accirding to [177]; however, in the earlier publication [45] the specific examples with an extremely large number of states - about 10^{16}, 10^{20} and even 10^{120} - are mentioned, successfully

---

[2] For parallel systems the global states (state of the whole system) and local states should be distinguished; formal definitions will be given later.

"model-checked"). Communication between concurrent processes in a system also complicates the verification.

The methods of analysis of Petri nets have been developed since the model has been designed by C.A. Petri [184]. They are deeply investigated in 3 main directions, complementary to each other: dynamic analysis (analysis by state space search), structural analysis and reduction methods [24, 26, 86, 172, 183]. Two main problems exist here. First is that in general case analysis of some important properties requires exponential time and memory. Second is that the methods working well for the objects like classical Petri nets usually cannot take into account the additional details important in the applications. For example, behavioral properties of an interpreted Petri net may differ from the properties of its underlying "pure" Petri net because of possible interaction between parts of the system via internal signals. So, a live Petri net may correspond to a non-live interpreted Petri net with the same structure, and vice versa.

Generally, we can say that modern CAD systems for discrete devices have insufficient possibilities for verification of parallel systems. They are usually unable to perform analysis tasks like deadlock detection, liveness and safeness checking. Such analysis is the designer's responsibility and is usually performed by means of simulation [71]. Of course, there is no guarantee of defects covering. In fact the CAD systems rarely provide any formal verification of designs, except syntax analysis (dynamic testing methods are also not sufficiently developed).

That is why further development of the analysis methods for Petri nets is important, such as studying the particular cases for which analysis tasks can be solved in polynomial time and working out the methods allowing to decrease time and memory amount in other cases, increasing in this way maximal size of the structures which can be practically analyzed. It is also important to develop the methods of analysis of the models which can describe the real-life devices and systems, such as interpreted Petri nets.

Discrete devices and systems, also parallel systems, may be synchronous or asynchronous. Methods of their analysis and synthesis differ for these two interpretations. Behavior of the asynchronous systems is in a sense more difficult; synchronous system can be considered as a particular case of asynchronous system and not vice versa. For a synchronous parallel automaton there always exists a behaviorally equivalent sequential automaton, but that is not always the case for an asynchronous parallel automaton [235]. However, some analysis tasks can be solved easier in case of the asynchronous systems.

Petri nets are asynchronous by their nature. So, methods of their analysis are "asynchronously oriented". We mentioned difficulties connected with applying methods of analysis of classical Petri nets to the interpreted Petri nets; similar difficulties appear when a Petri net with synchronous interpretation has to be analyzed. In both cases the main problem can be formulated as follows: an interpretation may forbid some of the possible evolutions of the underlying net. These forbidden evolutions are difficult to detect, and lack of them and may change remarkably the net properties (an unsafe net may become safe, some deadlocks may become unreachable etc.). The same is true for the FSM networks and their analysis. Generally, only two main properties are "resistant" with respect to interpretations: 1) if a Petri net is safe, all its interpretations are safe; 2) if an asynchronous system is deadlock-free, the corresponding synchronous system is also deadlock-free.

A system designer deals more often with asynchronous systems, although most of discrete devices are synchronous. Synchronousness is important at the low level, but a complex parallel system, even consisting of synchronous blocks, behaves asynchronously (a system of controllers of an automatized production line, for example, or a computer operating system). At the low level, asynchronous devices also find an important place, especially in the area of logical control (reactive embedded systems), because they are generally more quick.

This book is devoted to the analysis of asynchronous systems by means of reduced (selective) state space search.

1.2 Preliminary Remarks

Author's experience and inspiration in design of analysis methods for parallel algorithms have been formed during his work with two research teams: group of prof. A.D. Zakrevskij in the Institute of Engineering Cybernetics, Belorussian Academy of Sciences, Minsk, and group of prof. M. Adamski, University of Zielona Góra, Poland. During his work with the first team, the author's main interests were concentrated on optimization of sequent and parallel automata and analysis of sequent descriptions and α-nets (representing structure of parallel algorithms [242])[104, 106, 108, 138, 139]. Results of the researches had been summarized in the Ph.D. thesis [105]. Working with the second team, the author has been involved in developing the advanced methods of analysis of Petri nets [114, 118, 119, 228] and verification of specifications for parallel logical control systems described by FSM networks [115, 116], SFC [124] and Statecharts [113].

The two teams have developed similar methodologies of design of digital controllers implementing parallel logical control algorithms, both based on Petri nets and sequent descriptions. The Minsk team has developed a convenient formal description language for parallel control algorithms and has deeply investigated the problems of formal verification and state encoding of parallel automata [39, 40, 235, 243, 245, 249]; the Zielona Góra team has concentrated on sequent description of parallel algorithms and their implementation in FPGA and, generally, in programmable matrix structures [4, 12, 19, 35, 68, 144, 151, 199]. Both teams developed concepts of parallel algorithm and parallel automaton, and their results complement each other. In [6] selected results of both teams are presented.

In this book the conceptual apparatus of these methodologies is used. Formal definitions will be given in Chapter 2, but it is reasonable to explain informally some basic notions here.

A *parallel discrete system* is a dynamic system, for which a state is characterized by a vector of discrete (often Boolean) variables, and some of these variables may change independently of each other. It is evident, that every discrete device may be considered as a parallel system at "low enough" level. The state of whole parallel system is called its *global state*, to distinguish from the *local states*, specified by the components of the state vector. At a given moment a system can be in one global state, but in more than one local state. For example, in case of a Petri net its marking corresponds to the global state, and its places to the local states.

A *parallel automaton* [4, 151, 240] is an extension of the finite state machine, which may be in several (local) states simultaneously. Its underlying structure can be specified by a safe Petri net (usually an EFC-net) [252].

State space, or *reachability graph* of a parallel discrete systems is the set of its reachable global states and the direct transitions between them.

Logical control is a control performed by binary (logical) signals. Here signals flowing in both directions - from control system to controlled objects and vice versa - are logical. Logical control systems are widely used in industry, transport, communication networks, home automation and so on [249]. A *logical control algorithm* is a formal specification of logical control (behavior of a control system in interaction with the controlled objects). This algorithm is *parallel*, if some of its operations can be performed simultaneously. Parallel automation is a representation of a *parallel logical control algorithm* - one of the main objects under research in this book.

Dynamic analysis is an approach based on simulation of a system and extracting knowledge about its behavior from the state space. The "direct" method of dynamic analysis is constructing of the full state space. But it is evident, that sucvh method can be practically applied only to relatively small (and bounded) parallel systems. For unbounded Petri nets there is a trick (the famous "ω" [172, 183]), allowing to "pack" infinite state space into a finite graph; of course, some information is lost. For the systems, which are bounded but have "too many" reachable states, a wide range of methods is developed based on the idea of constructing reduced state spaces and extracting behavioral properties from them. Those methods are known as *lazy state space constructions* [85, 86] or *partial order reduction* [10, 44, 45, 77, 211]. Another family of methods which can be considered as belonging to the dynamic analysis approach are the *compression techniques*, representing the state spaces in a memory efficient manner, often (but not always) using BDD (see [34, 86, 89, 134, 150, 156, 167, 200, 217]). Such methods are beyond the scope of this book.

1.3   The Scope of the Book

Different aspects of analysis and verification of the parallel discrete systems, Petri nets and parallel logical control algorithms are discussed in hundreds of papers. The aim of this book is to present in systematic form the dynamic analysis approach to solving tasks of verification of Petri net and FSM network models and specifications, intended for implementation in embedded reactive systems. The dynamic analysis approach was selected here as the main strategy of system verification, because it allows not only deciding the behavioral properties, but also obtaining paths in state space, leading to the undesirable states or events, which is of much help for the design process. The Valmari's *stubborn set method*

[211, 212, 213, 215, 217], one of the most known methods of dynamic analysis of parallel systems, has been selected as the basic one. However, other approaches are also used in the book.

There exist good monographs describing theory and methodology of dynamic analysis of parallel systems (e.g. [77, 145, 187, 217, 221]). Contribution of our work in comparison to them is formed mostly by new results on properties and analysis of special classes of Petri nets, on combining different methods within the dynamic analysis approach, on net analysis by decomposition and, last but not least, on applying the approach to the detailed system descriptions such as interpreted Petri nets. One of the problems preventing wide practical use of dynamic analysis methods is that most of them is developed for too "abstract" state-transition systems, and usually they do not take into account some aspects of real-life parallel systems, important for their behavior. Dealing with this problem by adapting the dynamic analysis methods to the low-level descriptions is one of the central topics of the book.

To be brief: *the main subject of this book is behavior analysis of parallel discrete systems by means of dynamic methods with partial construction of state spaces.* The proposed methods are oriented to analysis of specifications and models of logical control systems.

There are 6 general directions of research, new results of which are presented in the book:

- studying additional possibilities of the known methods, such as stubborn set method, to obtain information about the properties of Petri nets. We show, that the stubborn set method can provide more information, than it was supposed before. Especially interesting results are obtained on analysis of the models corresponding to typical structures of parallel logical control algorithms - α-nets [109, 119, 124], nets with single-token initial marking [114] (Chapter 3);
- developing methods of net analysis with minimization of memory amount (on-the-fly reduction of reachability graphs [112, 118, 123, 131]); such methods can be useful, when memory is more critical parameter than time (Section 3.6);
- studying possibilities of simplifying net analysis by means of decomposition. A known method of analysis by decomposition was generalized for a class of cyclic nets [251]; an original approach to net decomposition for analysis has been developed [16, 107, 126, 127] (Chapter 4);
- developing methods of concurrent and distributed analysis of large nets (with the next motivation: it seems to be natural to analyze parallel structures in parallel way. Two variants of such analysis are considered: on common memory [130] and distributed [128]) (Sections 4.4, 4.5);
- developing algorithms of verification of the low-level, detailed specifications of embedded real-time systems, presented as interpreted Petri nets, Statecharts, SFC and FSM networks [14, 113, 117, 124]. A generalization of the stubborn set method has been considered, which can be easily concretized for many special parallel discrete models [121] (Section 3.2). Another approach,

used for analysis of Statecharts and FSM networks, is based on symbolic calculation of siphons in the modelling Petri nets [113, 116, 153] (Section 6.1). The improved algorithms of calculation of siphons and traps by means of solving logical expressions are applied for that purpose [228] (Chapter 5);

- verification of transformations and implementation of parallel automata specifications. Dynamic (testing) [120] and static (symbolic) [122] approaches to such verification are considered (Section 6.2).

Besides, some new theoretical results and algorithms are presented, obtained as the by-products of our main research. They are related to such topics as behavioral properties of Petri nets [110, 119] (subsection 3.5.1), cycles in oriented graphs [111] (Appendix B), and generating prime implicants of Boolean functions [32, 33, 228] (Appendix D).

2. Main Notions, Problems and Methods

2.1 Models and Specifications of Parallel Discrete Systems

This section contains definitions of main models used in the book and their properties. Other necessary definitions will appear in the text when correspondent notions will be introduced; they, unlike the definitions in this section, will be numbered.

2.1.1 Parallel Discrete Systems (General Definition)

By a *parallel discrete system* we mean a system which (global) state M is described by a vector of discrete variables and can be changed by firing (execution) of the *transitions*; every transition has a necessary condition of firing being a binary function defined on M (when it is satisfied, the transition is *enabled*). A transition firing changes values of some elements of M. Several transitions can be enabled in the same state M. Set of all enabled transitions at M is denoted as $enabled(M)$.

If transition t is enabled in M and its firing transforms the state into M', that is denoted by MtM'. This denotation and the notion of enabled transition can be generalized for *firing sequences* (the sequential firing of the transitions, such that each transition is enabled in the state created by firing of the previous transition). A state M' that can be reached from M by a firing sequence $\sigma = t_1 t_2 ... t_n$ is called *reachable* from M; we write $M \sigma M'$; $|\sigma| = n$; $t_i \in \sigma$ $(1 \leq i \leq n)$. For a state M, $[M\rangle$ denotes the set of all states reachable from M.

A transition is *live*, if there is a reachable state in which it is enabled; otherwise it is *dead*. A state in which no transitions are enabled is called a (global) *deadlock*.

A (full) *reachability graph* is a graph $G = (V, E)$ representing state space of a system. $V = [M_0\rangle$; $d = (M, M') \in E \Leftrightarrow MtM'$ (then t marks d). A *strongly connected component* (SCC) of a reachability graph is its maximal strongly connected subgraph. A *terminal component* of a graph G is its SCC such that each edge which starts in the component also ends in it [74, 217]. A system is *cyclic*, if its reachabilty graph is strongly connected.

A. Karatkevich: Dynamic Analysis of Petri Net-Based Discrete Sys., LNCIS 356, pp. 9–26, 2007.
springerlink.com

All models described below are concretizations of this general model. The notions presented above will be used for those models.

2.1.2 Petri Nets and Their Extensions

Petri Nets

An (ordinary) *Petri net* [172, 188] is a triple $\Sigma = (P, T, F)$ where:

P is a finite nonempty set of *places*;
T is a finite nonempty set of *transitions*;
$P \cap T = \emptyset$;
F is a set of *arcs*: $F \subseteq (P \times T) \cup (T \times P)$.

Sets of input and output transitions of place p are defined respectively as follows:

$$^\bullet p = \{t \in T : (t, p) \in F\},$$
$$p^\bullet = \{t \in T : (p, t) \in F\}.$$

Sets of input and output places of transition t are always nonempty and are defined respectively as follows:

$$^\bullet t = \{p \in P : (p, t) \in F\},$$
$$t^\bullet = \{p \in P : (t, p) \in F\}.$$

These notions can be generalized for sets of places (transitions); for example, $^\bullet T$ is the set of all input places of the transitions belonging to T.

Graphically a Petri net is represented as a bipartite oriented graph, which nodes correspond to places and transitions, and arcs are going from transitions to their output places and from places to their output transitions (Fig. 2.1). A net is said to be *connected* (*strongly connected*), if its graph is connected (strongly connected).

A state of a Petri net, called a *marking*, and is defined as a function $M : P \to \mathbb{N}$. It can be considered as a number of tokens situated in the net places (p contains $M(p)$ tokens). $M(P')$, where $P' \subseteq P$, denotes $\sum_{p \in P'} M(p)$. When a place or set of places contain token(s), it is *marked*. A marking M is *safe*, if $\forall p \in P(M(p) \leq 1)$, otherwise it is *unsafe*. We will specify the safe markings as the sets of marked places. A transition t is *enabled* and can *fire* (be executed), if $\forall p \in{}^\bullet t(M(p) > 0)$. Transition firing removes one token from each input place and adds one token to each output place. A marking can be changed only by transition firing. The initial marking M_0 is usually specified (then the Petri net is represented as a tuple $\Sigma = (P, T, F, M_0)$). $(M \geq M') \Leftrightarrow \forall p \in P(M(p) \geq M'(p))$; $(M > M') \Leftrightarrow (M \geq M') \wedge (\exists p \in PM(p) > M'(p))$.

Transitions t, t' are *parallel*, if $\exists M \in [M_0\rangle\{t, t'\} \subseteq enabled(M)$ and $^\bullet t \cap {}^\bullet t' = \emptyset$. Transitions t, t' such that $^\bullet t \cap {}^\bullet t' \neq \emptyset$ are *in conflict*.

It is supposed, that transitions of a Petri net fire one by one; however, for the simulation purposes a notion of *step*step (of simulation) (of concurrent simulation) is sometimes useful. The *step*step (of simulation) is a set Δ of enabled

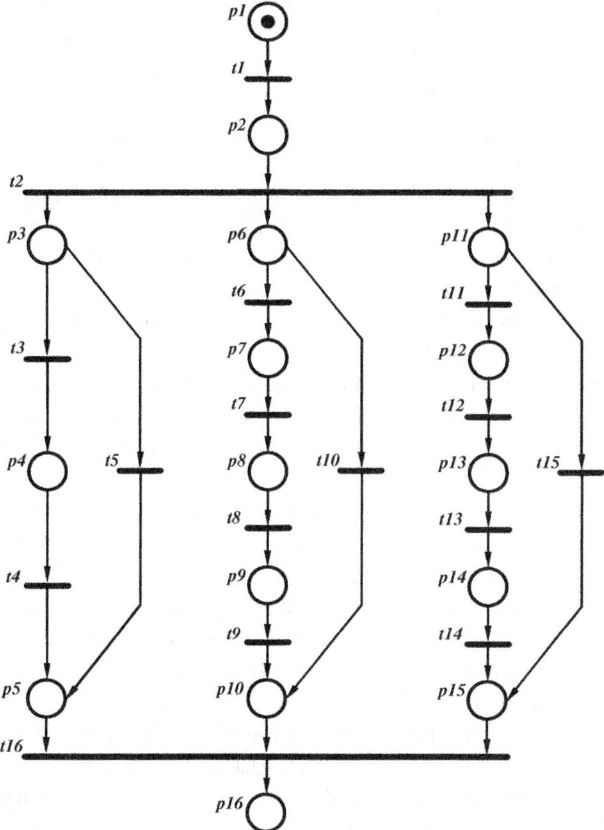

Fig. 2.1. A Petri net

mutually independent transitions [97] (see Definition 3.1); transitions belonging to a step can fire in any order with the same marking resulting. Notion of firing sequence can be generalized for *step sequences*. A *linearization* of a step is a sequence of all transitions belonging to it in any order. A linearization of a step sequence is the concatenation of linearizations of all steps in the sequence.

A net is *live*, if for every reachable marking every transition of the net is live. A net is *quasi-live*, if for the initial marking every transition of the net is live. A net is *bounded* (n-bounded), if $\exists n : \forall p \in P \, \forall M \in [M_0\rangle (M(s) \leq n)$ (there is an upper bound of number of tokens for all the net places in all reachable markings). A net is *safe* (1-bounded), if $\forall p \in P \, \forall M \in [M_0\rangle (M(s) \leq 1)$. A live and safe (LS) net is often called a *well-formed* net. A net is *reversible*, if $\forall M \in [M_0\rangle \, M_0 \in [M\rangle$ (it can return to the initial marking from any reachable marking). A net is *conservative*, if the number of tokens in all reachable markings is the same (or, in more general formulation, if such non-zero coefficients can be associated with the places, that the weighted sum of tokens is constant for all reachable markings).

A Petri net $\Sigma' = (P', T', F')$ is a *subnet* of Petri net $\Sigma = (P, T, F)$, if:

$$T' \subset T;$$
$$P' = {}^\bullet T' \cup T'^\bullet; \tag{2.1}$$
$$F' = F \cap ((P' \times T') \cup (T' \times P')).$$

It follows that a subnet of the given net is completely specified by the set T'. We will use set-theoretical operations for the subnets, interpreted as follows: $\Sigma_1 \odot \Sigma_2 = \Sigma_3$, where $T_3 = T_1 \odot T_2$, $\odot$ is a set-theoretical operation, and Σ_3 is specified by T_3 according to (2.1).

A *projection* of marking M of a net Σ on its subnet Σ' is marking M' of Σ' such that $\forall p \in P' \; M'(p) = M(p)$.

A net belongs to the class SM (*State Machine*), if $\forall t \in T : |{}^\bullet t| = |t^\bullet| = 1$. A net $\Sigma' = (P', T', F')$ belonging to the class SM, such that for a net $\Sigma = (P, T, F)$ $P' \subseteq P$, $T' \subseteq T$, $F' = F \cap ((P' \times T') \cup (T' \times P'))$, is called an *SM-component* of Σ.

A net belongs to the class EFC (*Extended Free Choice*) [55], if $\forall p \in P : (t \in T, t' \in T, p \in {}^\bullet t, p \in {}^\bullet t') => ({}^\bullet t = {}^\bullet t')$. A set of transitions of EFC-net with the same input sets of places is a *cluster*; clusters specify a partition of the set of places and of the set of transitions. An EFC-net is called an α-*net* [249], if in the initial marking it contains only one token. For every element $x \in P \cup T$, $[x]$ denotes the smallest set containing x which includes $p^\bullet$ for every place $p \in [x]$ and ${}^\bullet t$ for every transition $t \in [x]$. The set $[x]$ is called a *cluster* of Σ [54] (the set of all clusters of Σ constitutes a partition of $P \cup T$; for every cluster c of an EFC-net, every place $p \in c$ and every transition $t \in c$: $(p, t) \in F$. All transitions in a cluster an EFC-net can be enabled or disabled only simultaneously; so, a cluster of an EFC-nets is *enabled*, if all transitions belonging to it are enabled; otherwise it is *disabled*). A *free choice* net is an EFC-net with the property that no cluster has simultaneously more than one transition and more than one place.

A *siphon* is a set of places such that every transition which outputs to one of the places in the siphon also inputs from one of these places. This means that once all of the places in the siphon become unmarked, the entire set of places will always be unmarked; no transition can place a token in the siphon because there is no token in the siphon to enable a transition which outputs to a place in the siphon [183] (Fig. 2.2). Formally, a nonempty subset of places D of a net Σ is a siphon if ${}^\bullet D \subseteq D^\bullet$.

A *trap* is a set of places such that every transition which inputs from one of these places also outputs to one of these places. This means that once any of the places in a trap has a token there will always be a token in one of the places of the trap. Firing of transitions may move the tokens between places but cannot remove all tokens from a trap [183] (Fig. 2.3). Formally, a nonempty subset of places U of a net Σ is a trap if $U^\bullet \subseteq {}^\bullet U$.

Union of two siphons (traps) is again a siphon (trap). A siphon (trap) is called a *basic siphon* (*basic trap*) if it cannot be represented as a union of other siphons (traps). So all the siphons (traps) in a Petri net can be generated by the union of some basic siphons (traps). A siphon (trap) is said to be *minimal* if it does

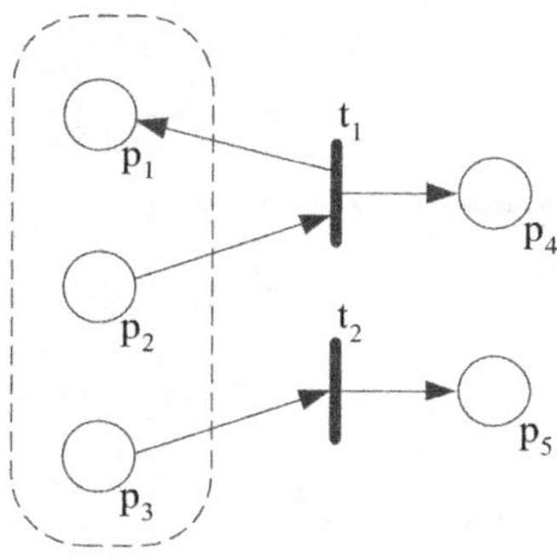

Fig. 2.2. Example of a siphon $D = \{p_1, p_2, p_3\}$, $^\bullet D = \{t_1\}$, $D^\bullet = \{t_1, t_2\}$

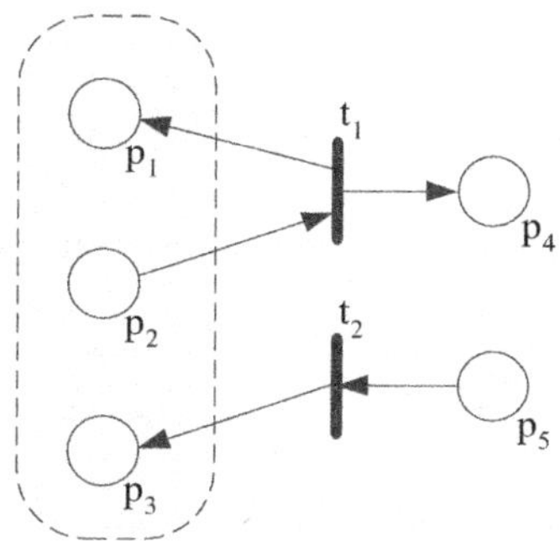

Fig. 2.3. Example of a trap $U = \{p_1, p_2, p_3\}$, $U^\bullet = \{t_1\}$, $^\bullet U = \{t_1, t_2\}$

not contain any other siphon (trap). Minimal siphons (traps) are basic siphons (traps), but not all basic siphons (traps) are minimal.

The Petri net shown in Fig. 2.4 contains 10 siphons and 10 traps (excluding the entire set of places, which is both a siphon and a trap). In Table 2.1 all those siphons and traps are listed; minimal siphons and traps are marked.

For a set $U \subseteq P$, $\chi[U]$ denotes the characteristic vector of U with respect to P.

The *incidence matrix* $C : P \times T \rightarrow \{-1, 0, 1\}$ of Σ is defined by $C(-, t) = \chi[t^\bullet] - \chi[^\bullet t]$.

Petri Nets with Multiple Arcs

A *Petri net with multiple (weighted) arcs* [172, 183] is a Petri net with a weight function defined on the set of arcs: $W : F \rightarrow \{1, 2, 3, ...\}$. For such nets:

1. A transition t is enabled, if each input place p of t is marked with at least $w(p, t)$ tokens, where $w(p, t)$ is the weight of the arc from p to t.
2. A firing of an enabled transition t removes $w(p, t)$ tokens from each input place p of t, and adds $w(t, p)$ tokens to each output place p of t, where $w(t, p)$ is the weight of the arc from t to p.

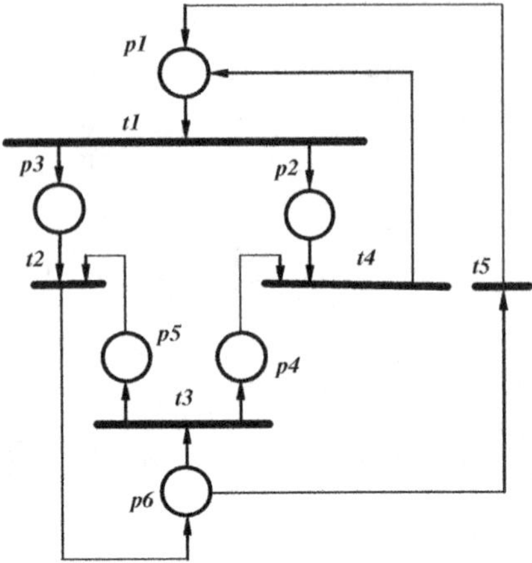

Fig. 2.4. A Petri net (taken from [140])

Table 2.1. Sets of siphons and traps of the Petri net shown in Fig. 2.4

siphon	minimal siphon	trap	minimal trap
$\{p_5, p_6\}$	+	$\{p_1, p_2\}$	+
$\{p_4, p_5, p_6\}$	-	$\{p_1, p_2, p_5, p_6\}$	-
$\{p_1, p_2, p_3, p_6\}$	+	$\{p_1, p_2, p_4\}$	-
$\{p_1, p_2, p_5, p_6\}$	-	$\{p_1, p_2, p_4, p_6\}$	-
$\{p_1, p_2, p_3, p_5, p_6\}$	-	$\{p_1, p_2, p_3, p_5, p_6\}$	-
$\{p_1, p_3, p_4, p_6\}$	+	$\{p_1, p_2, p_4, p_5, p_6\}$	-
$\{p_1, p_2, p_3, p_4, p_6\}$	-	$\{p_1, p_3, p_4, p_6\}$	+
$\{p_1, p_4, p_5, p_6\}$	-	$\{p_1, p_2, p_3, p_4, p_6\}$	-
$\{p_1, p_2, p_4, p_5, p_6\}$	-	$\{p_1, p_3, p_4, p_5, p_6\}$	-
$\{p_1, p_3, p_4, p_5, p_6\}$	-	$\{p_1, p_3, p_5, p_6\}$	+

An example of net with multiple arcs see in Fig. 6.5c.

Interpreted Petri Nets

An *interpreted Petri net* in narrow sense is a Petri net enhanced with possibilities of information exchange (by means of binary signals) with the outer world, and with exact determination when an enabled transition fires. Interpreted Petri nets are mostly used as models of parallel discrete devices, such as logical controllers [7, 12, 13, 14, 19, 20, 34, 51, 67, 103, 218, 225, 249].

Interpreted Petri nets can be, like the FSMs, of Mealy or Moore type. An interpreted net of Mealy type is a Petri net such that a *sequent* is associated to every transition t_i of it. A *sequent* [4, 237, 249] is an expression of the form $\varphi_i \vdash \psi_i$, where φ_i is a Boolean function, and ψ_i is an elementary conjunction.

An enabled transition of such net fires, when in the corresponding sequent $\varphi_i = 1$. Firing of the transition leads to assigning to all variables occurring in ψ_i the values, for which $\psi_i = 1$. The Boolean variables occurring in the sequents are divided into 3 sets: X - input variables, occurring only in left parts of the sequents; Y - output variables, occurring only in the right parts; and Z - internal (shared) variables, occurring in both (often $Z = \emptyset$). These variables are the signals of communication between the net and outer world (and, when $Z \neq \emptyset$, also of communication between processes inside the net. Values of input variables depend on the outer world; values of the internal and output variables can be changed only by the transition firing. An example of interpreted Petri net of Mealy type is shown in Fig. 6.1. *Underlying net* of an interpreted Petri net is the classical Petri net with the same sets of places, transitions and arcs (it can be obtained from an interpreted Petri net by removing all information about the signals). The underlying net for the net from Fig. 6.1 is the net shown in Fig. 3.1.

An interpreted net of Moore type is such Petri net that a *condition* (a Boolean function of the input variables, belonging to the set X) φ_i is associated to every transition t_i of it, and a set of Boolean variables (maybe empty) being a subset of the set of output variables Y is associated to every place. An enabled transition of such net fires, when the corresponding condition $\varphi_i = 1$. At every marking the output variables, which are associated with the places having tokens, have value 1, other output variables have value 0. Values of the input variables depend on the outer world; so, input and output variables provide an interface between a Petri net and outer world. An example of interpreted Petri net of Moore type is shown in Fig. 6.15.

Input and output variables of an interpreted Petri net can be understood, among others, as input and output signals of a logical controller, and the whole net as a specification of parallel logical control algorithm.

Nets with Inhibitor Arcs

Modelling power of Petri nets is limited by lack of possibility of zero-test (there is no possibility to check whether a place is empty). That's why for modelling systems and describing control algorithms the extended Petri net model known as *Petri nets with inhibitor arcs* [43, 183] is often used. An *inhibitor arc*, leading from a place to a transition, unlike the "traditional" arcs, disables the transition, if the place has a token.

An inhibitor arc is denoted graphically by an arc with a small circle (see Fig. 6.4). The set of places from which the inhibitor arcs lead to transition t is denoted by $(^\bullet t)^{(inh)}$.

Nets with Priorities

A Petri net with (static) *priorities* [30] is a Petri net such that a *priority relation* $\rho \subseteq T \times T$ is defined on the set of transitions; $(t_1, t_2) \in \rho$ means, that transition t_2 has higher priority than t_1. A transition can be enabled in a marking, only when no transition with higher priority is enabled in this marking. A transition which is disabled but has no empty input place is *active*

An example of a net with priorities and its reachability graph is shown in Fig. 6.5.

2.1.3   Parallel and Sequent Automata

Parallel Automata

Parallel automaton [4, 41, 186, 201, 239, 240, 249] is a model equivalent to the interpreted Petri net (with the restriction: the underlying Petri net should be a safe α-net). Such automaton is considered as a dynamic system which can be in several *local* states simultaneously. A marking of a Petri net corresponds to a *global* state of a parallel automaton; a place corresponds to a local state (*active*, when the corresponding place is marked, *passive* otherwise). A set of simultaneously active local states specifies a *global state*. Any two local states which can be active simultaneously are called *parallel*.

A parallel automaton is described by set of transitions of the form $\mu - \varphi \vdash \psi \to \nu$, where φ and ψ are elementary conjunctions of Boolean variables that define the condition of transition and the output signals respectively, μ and ν are labels that represent the sets of local states of the parallel automaton [186, 240, 247, 249]. Every such transition should be understood as follows. If a global state of the parallel automaton contains all local states from μ and $\varphi = 1$, then the transition is executed (fired), which yields a new global state that differs from the previous one by containing local states from ν instead of those from μ. The values of output variables in this case are set to be such that $\psi = 1$.

A local state p is *reachable*, if such global state is reachable that p is active. This notion relates to parallel automata, Statecharts and FSM networks (defined below).

Sequent Automata and State Encoding

A *sequent automaton* [237, 249] is a system S of sequents $s_i = \varphi_i \vdash \psi_i$, defining the "cause-effect" relation between an event represented by Boolean function φ_i and an event represented by conjunction term ψ_i [250]. Set of the arguments of all functions in the system is divided into 3 sets, as for the interpreted Petri nets. If at a moment $\varphi_i = 1$ (the sequent is enabled), then conjunction ψ_i will attain value 1 (after an arbitrary delay, in case of asynchronous interpretation; at the next moment of discrete time, in case of synchronous interpretation); such change of the values of corresponding variables is called *sequent firing*. The internal and output variables are *inertial* (keep their values, when no sequent firing changes them).

A *simple sequent automaton* is defined as a system of *simple* sequents - the sequents where both parts φ and ψ are the elementary conjunctions.

Sequents s_i and s_j are called *parallel* if they could be enabled simultaneously. A sequent automaton is *consistent*, if for any parallel sequents s_i and s_j relation $\psi_i \wedge \psi_j \not\equiv 0$ holds [250].

A safe interpreted Petri net or a parallel automaton can be converted into behaviorally equivalent system of sequents. The main operation of such conversion is *state encoding* (or *state assignment*) [40, 41, 186, 245, 247]. In state assignment of a parallel automaton, local states are encoded by ternary vectors (elements of which can have values "0", "1" or "-") in the space of introduced internal (coding) variables, non-orthogonal vectors being assigned to parallel states (non-parallel states are usually, but not always, coded by orthogonal vectors) [246]. The orthogonality of ternary vectors means existence of a component having opposite values (0 and 1) in these vectors. After state encoding is performed, a sequent automaton is obtained from the parallel automaton in the following way: for each transition $\mu - \varphi \vdash \psi \rightarrow \nu$ a sequent $\varphi f_z \vdash k_z \psi$ is obtained. Here $f_z = \bigwedge_{p \in \mu} Qp$, $k_z = \bigwedge_{p \in \nu} Qp$, where Qp is the code of local state p, interpreted as the conjunction of coding variables [40, 249].

2.1.4 Sequential Function Charts

Sequential Function Charts (SFC) is a graphical language for logical control systems specification, allowing parallel branches description. It is one of the IEC (International Electrotechnical Commission) programming languages for industrial control systems [159].

An SFC consists of steps and transitions (Fig. 2.5a). A *step* of SFC represents a particular state of a system being controlled. It may be active or not. A step can be associated with one or more actions. A *transition* is associated with a condition; if all steps before a transition are active and the condition is true, the transition occurs. Then all steps before the transition become inactive, and all steps after the transition become active. Relations "before" and "after" are specified by the lines at SFC diagram (we do not explain in detail the SFC graphical specification; details can be found in [159]). Initially only one step is active.

An SFC is *unsafe* if a step can be activated when it is already active (Fig. 2.5b); this situation would mean an unpredictable behavior of the controller. Another case of defected SFC (unreachable) is when there are steps that never can be activated (Fig. 2.5c; all examples in Fig. 2.5 are taken from [159]). SFC may have one or more *terminal steps*. From any reachable global state, a global state with an active terminal step should be reachable. If there are no terminal steps, the initial state should be reachable from any reachable state.

2.1.5 Statecharts

Statecharts [83, 88], like SFC, is a graphical language used for specification of control systems. We use the restricted model, not including such elements of classical Statecharts, as negated trigger events, timeout events, and assignment to variables. The definitions are based on [46, 149, 151, 173].

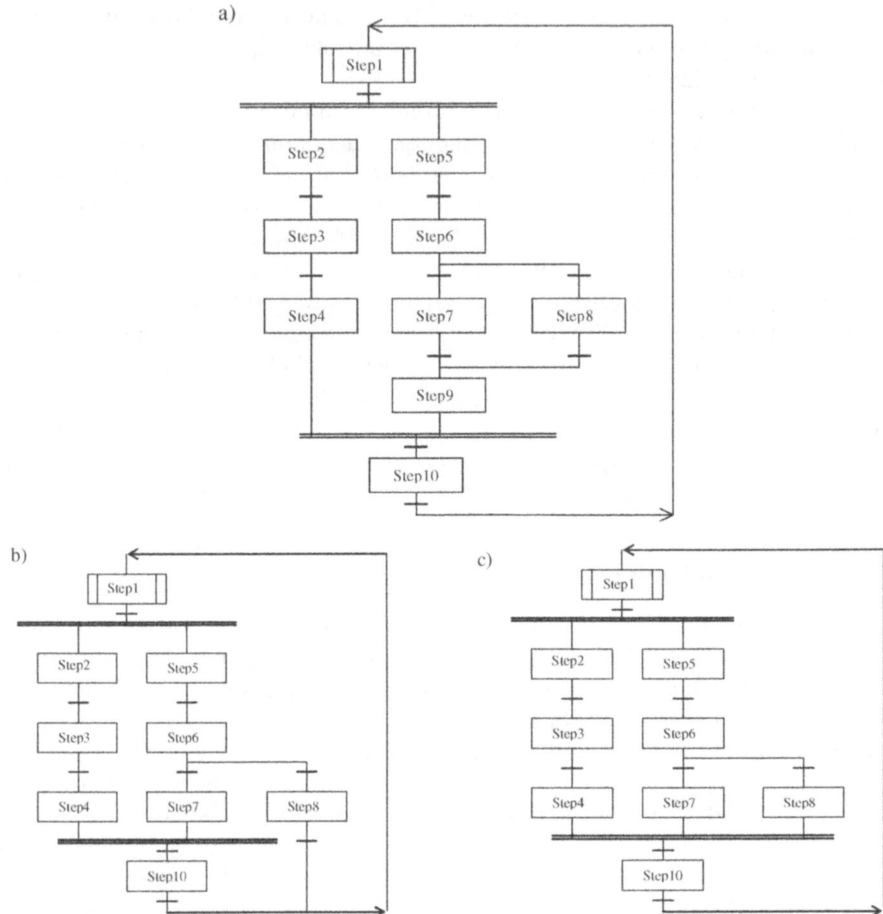

Fig. 2.5. The examples of SFC

Syntax of Statecharts

A *Statechart* is an 11-tuple consisting of the following elements:

$$(P, hrc, type, default, history, \Gamma, T, out, in, tlabel, saction),$$

where:

1. P is the finite non-empty set of states.
2. $hrc\colon P \to 2^P$ is the hierarchy function, which assigns to every state $p \in P$ the set of *immediate substates* of p. Transitive closure of the relation of being an immediate substate is the relation of being a *substate* (the reversed relation specifies being a *superstate*). It is supposed that the graph specified by the relation of being an immediate substate is a tree (no two states are the substates of each other; no state is an immediate substate of more than one

state; there is a state *root* for which all other states in the diagram are the substates). $(parent(p) = p') \Leftrightarrow p \in hrc(p')$.

3. *type*: $P \rightarrow \{AND, OR\}$ is the *state-type function*.
4. *default*: $P \rightarrow P$ is the *default function* (which for every state $p \in P$ such that $hrc(p) \neq \emptyset$ and $type(p) = OR$ assigns default state of hrc, and is undefined otherwise.).
5. *history*: $P \rightarrow \{true, false\}$ is the Boolean *history function*.
6. Γ is the finite set of events; $\Gamma = I \cup Z$; $I \cap Z = \emptyset$ (I - the set of *external events*, Z - the set of *internal events*).
7. T is the finite set of *transitions*.
8. *out*: $T \rightarrow P\backslash\{root\}$ is the *source function*: $out(t) = p$ if transition t originates from state p.
9. *in*: $T \rightarrow P\backslash\{root\}$ is the *target function*: $in(t) = p$ if transition t ends in state p.
10. *tlabel*: $T \rightarrow 2^{\Gamma} \times 2^{Z}$ is the *transition labelling function*. The first component of *tlabel* is $trigger(t) \subseteq \Gamma$, the second is called *transition action* and is denoted as $taction(t) \subseteq Z$.
11. *saction*: $P \rightarrow 2^{\Gamma}$ is the *state labelling function*, defining the set of events which are available when the state is active.

For every transition $t \in T$, the following predicate holds: $parent(out(t)) = parent(in(t)) = p$ with $type(p) = OR$.

In Fig. 2.6 an example of Statechart is shown (taken from [151], with minor changes. The transitions are described as $t : trigger(t)/taction(t)$, or as $t : trigger(t)$, if $taction(t) = \emptyset$).

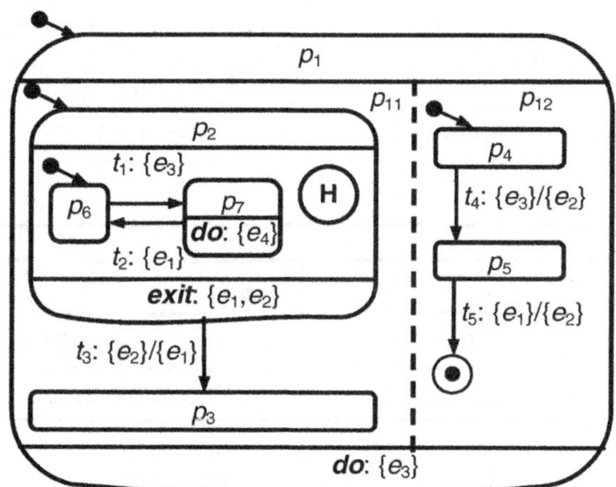

Fig. 2.6. An example of Statechart

Behavior Specified by Statecharts

As far as we consider Statecharts as a description of system behavior, it is necessary to specify their behavioral interpretation. So, a *global state* M is defined as a set of *active states*, and function $M(p)$ is defined such that $M(p) = 1$ if and only if p is active in M. Initial global state M_0 is defined by the default function: $M_0(root) = 1$; $\forall p \in (P \backslash \{root\}) : M_0(p) = 1 \Leftrightarrow (M_0(parent(p)) = 1 \wedge ((type(parent(p)) = AND) \vee (default(parent(p)) = p)))$.

A transition $t \in T$ is *active* and can *fire* in M, if and only if $M(out(t)) = 1$ and all the events from $trigger(t)$ *occur*. Firing of a transition t changes the global state: $out(t)$ and all its substates become inactive; $in(t)$ becomes active, and its substates become active similarly as it is defined for the initial state (according to the default function). The exception is a state with history; when it becomes active, its immediate substate last being active becomes active again. Occurrence of the events from I depends only on the external world and can happen asynchronously. An event $e \in Z$ occurs, if and only if a transition t is firing such that $e \in trigger(t)$ (*dynamic event*) or state p is active such that $e \in saction(p)$ (*static event*). An active transition t fires immediately, if it is not in conflict with another active transition; in case of a conflict only one of the conflicting transitions can fire. An event e is *consumed* by firing of transition t such that $e \in trigger(t)$ (becomes unavailable; if it is necessary to keep the event available, it can be generated again by t). If an event had not been consumed by a transition immediately, it becomes unavailable. But although an event is available only during an instant of time, for analysis purposes we have to complement the notion of global state by the notion of *available events* (if e is available, $M(e) = 1$). A transition $t \in T$ is *enabled* if $M(out(t)) = 1$ and all the events from Z occur. Intuitively, a transition is enabled if it is active or if it lacks only the external events to be active. Otherwise it is *disabled*.

A *global deadlock* of a Statechart is such a global state that no transition can fire, independently of the occurring external events. A *local deadlock* is a set U_d of the active states such that no transitions originating from them can become active, because each of them waits for the events which cannot be generated without firing of other transitions originating from the states belonging to U_d. A simple example of a deadlock is provided by Fig. 2.7 [151].

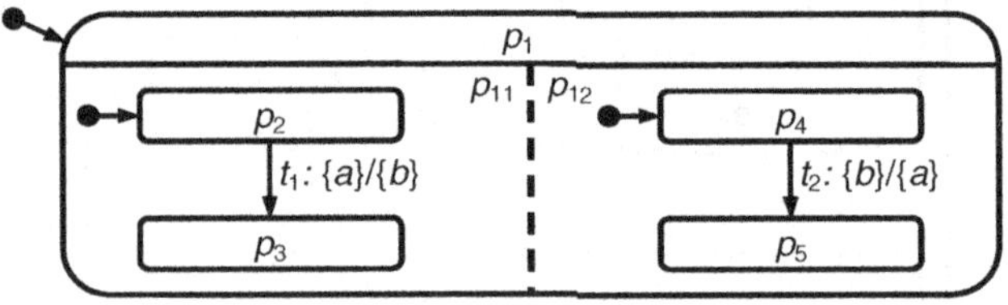

Fig. 2.7. Example of a deadlock

2.1.6 Finite State Machines and FSM Networks

A *finite state machine* (FSM) is usually defined [73, 75, 91] as a 5-tuple $(S, I, O, \lambda, \delta)$, where S is a finite set of states, I is a finite input alphabet, O is a finite output alphabet, $\lambda : S \times I \to S$ is a state transition function, $\delta : S \times I \to O$

is an output function. However, we will use an alternative, event-based definition [157], which is more convenient for our purposes.

An FSM is a 4-tuple $A = (I, O, P, T)$, where I is a set of input events, O is a set of output events, P is a set of states, T is a set of transitions. An event is a named variable that is either *present* or *absent*. Each transition $t \in T$ is $t = (p_s, guard, p_d)$, where $p_s \in P$ is the source state, $guard \subseteq I$, $p_d \in P$ is the destination state. A set of events $O_i \subseteq O$ is associated with every state (Moore FSM) or every transition (Mealy FSM). A transition is triggered when the FSM is in the transition's source state and all the events belonging to the guard are present. Then the FSM goes to the destination state. An event $e \in O$ is present, if and only if the FSM is in such state p_i that $e \in O_i$ (Moore FSM) or transition t_i such that $e \in O_i$ is being executed (Mealy FSM).

An FSM network $N = \{A^1, A^2, ..., A^n\}$ is a set of FSMs, in which the situation is possible such that $I^i \cap O^j \neq \emptyset$. A *global state* of an FSM network is an n-tuple $M = (p_p^1, p_r^2, ..., p_m^n)$, where p_j^i is a state of automaton A^i. A global state M_k is reachable from M_l, if and only if for M_l there exists such input sequence (sequence of input events), that after applying it the local states can form M_k, according to the transition guards. We assume *asynchronous* interpretation of FSM networks: state of a component automaton changes with an arbitrary delay. A *deadlock* is a situation where two or more FSMs cannot proceed (are *deadlocked*) because each is waiting for events from the others.

FSM networks can be considered as a particular case of Statecharts - the Statechart diagrams such that

$$
\begin{aligned}
&type(root) = AND, \\
&\forall p \in hrc(root) : type(p) = OR, \\
&\forall p \in hrc(hrc(root)) : hrc(p) = \emptyset.
\end{aligned}
\tag{2.2}
$$

It is easy to see, that a Statechart satisfying (2.2) describes communicating automata without hierarchy and is equivalent to an FSM network, as defined above.

Some other models, such as Petri net models, can be converted into FSM networks, if their underlying nets are covered by SM-components (which is usual for logical control applications) [172, 249].

Description of FSM networks by means of Statecharts is used in this book. The top indexes at states and transitions are used to denote to which FSM they belong.

2.2 The Tasks of Analysis

Analysis of parallel models such as Petri nets is important in most of their applications. And, however the applications may differ very much, there are some Petri net properties, which are typical for the nets modelling the "correct" systems. For example, usually the modelled or specified systems are supposed to be finite (especially in the digital design applications). Then the corresponding Petri nets should be bounded. Practically in most of cases the condition is stronger - the nets should be safe (then a global state is described by a binary vector).

Another typical condition is liveness (or, in some cases, quasi-liveness). If a net is not live, then the system has fragments which can become "dead" (if it is not even quasi-live, then there are fragments, which are initially "dead" and hence redundant). Usually it means, that something is wrong with the system. A particular case of non-liveness is existence of reachable deadlocks: if the system is live, then no deadlocks exist; if it should have one or more terminal states, then the reachable deadlocks should exactly correspond to the specified terminal states.

From the above two very important Petri net analysis tasks can be formulated. One of them is deciding "well-formedness" (liveness and safeness together) of a net (for certain classes of nets there are methods, checking whether the net satisfies both properties, liveness and safeness (or boundedness), or not [54, 104, 105, 124, 135, 243, 245, 249]). Another is deadlock detection. Probably most of developed methods of Petri net analysis are dedicated to solving those tasks.

Other well-known tasks of Petri net analysis are deciding behavioral properties of the net, such as persistence and reversibility, and also the reachability and coverability analysis [74, 86, 172, 217]. Most of them can be applied to analysis of parallel algorithms. In the detailed theory of logical control algorithms described in [39, 40, 235, 243, 245, 247, 248, 249] an algorithm of the type examined is termed correct if it is compact, restorable, self-consistent, noncontradictory, and stable. First three of those properties are satisfied if the underlying Petri net is well-formed; two others are related to the level of interpreted Petri nets and can be easily decided, if the relation of parallelism between the transitions is known.

Localizing of the faults localization would be also useful for verification of algorithms. But for some faults related with parallel branches interaction localization in terms of the algorithm units (or Petri net places and transitions) seems to be impossible That is the case, for example, when a Petri net is not live but there are no reachable deadlocks. Then the most acceptable way of localization is obtaining a path (a firing sequence) leading to a "wrong" situation.

One else important area of Petri net analysis is structural analysis. Structural properties, unlike the behavioral ones, do not depend on the initial markings (there exist, however, the properties which can be considered as "intermediate" between behavioral and structural ones, such as *structural liveness* and *structural boundedness* [172]; those properties are formulated as "exists such initial marking, that ... " or "for any initial marking ... "). Relation between structural and behavioral properties of Petri nets is illustrated e.g. by the well-known results on free choice nets [82] and EFC-nets [29]. Also, structural analysis, such as detection of SM-components, is important for some synthesis tasks [8, 249]. Usually the tasks of structural analysis can be reduced to calculation of siphons and traps [172, 183].

In this book we concentrate mostly on the analysis of behavioral properties, paying special attention to localization of faults. Deadlock detection is selected as the "basic" analysis task; other tasks, as deciding liveness and safeness, are also considered. Chapter 5 is the only chapter dedicated to structural analysis (calculation of siphons and traps).

2.3 The Methods of Analysis

The methods using lazy state space constructions are reviewed in Section 3.1, some linear- and logical-algebraic methods - in Chapter 5. Here we present a short review of other methods of parallel discrete system analysis.

The most known method of Petri net analysis is the **coverability graph (*coverability tree*) method** [172, 183, 217]. We present it in the variant of *coverability graph*. For given Petri net $\Sigma = (P, T, F, M_0)$:

Algorithm 2.1

1. Introduce M_0 as a node and tag it "new".
2. While "new" markings exist, do the following:
   a) Select a new marking M.
   b) If no transitions are enabled in M, tag M "deadlock".
   c) While there exist enabled transitions in M, do the following for each enabled transition t in M :
      i. Obtain the marking M' that results from firing t from M.
      ii. If there is no node M', introduce M' as a node and tag M' "new".
      iii. Introduce an arc (M, M'), labelled by t.
      iv. If on a path from M_0 to M there exists a marking M'' such that $M' > M''$, then replace $M'(p)$ by ω for each p such that $M'(p) > M''(p)$.
   d) Remove label "new" from M.
3. The end.

In the case of bounded nets the coverability graph directly represents the state space and thus completely describes the behavior. In the case of unbounded nets (infinite state space) coverability graph is still finite (see the example, taken from [183] - Fig. 2.3). For the unbounded nets construction of coverability graphs in general case is not enough neither for reachability analysis, nor for liveness checking (examples see in [172, 183]). The coverability graph method demonstrates the state explosion at full extend.

The idea of **net reduction** methods [24, 25, 26, 172, 235, 249] is simplifying the net in a way preserving the properties we want to decide (usually liveness and safeness). As an example a set of transformations preserving liveness, safeness and boundedness is depicted in Fig. 2.9 (taken from [172]). Usually the net reduction methods have to be used in combination with other methods; however, in some cases reduction is enough to decide if a given net is well-formed. In [235] the next reduction rules are proposed:

Elimination of Loops: transition t_i can be removed from the net, if $^\bullet t_i = t_i^\bullet$ and there is another transition $t_j \in T$ such that $^\bullet t_i \subseteq {}^\bullet t_j$ or $t_i^\bullet \subseteq t_j^\bullet$.

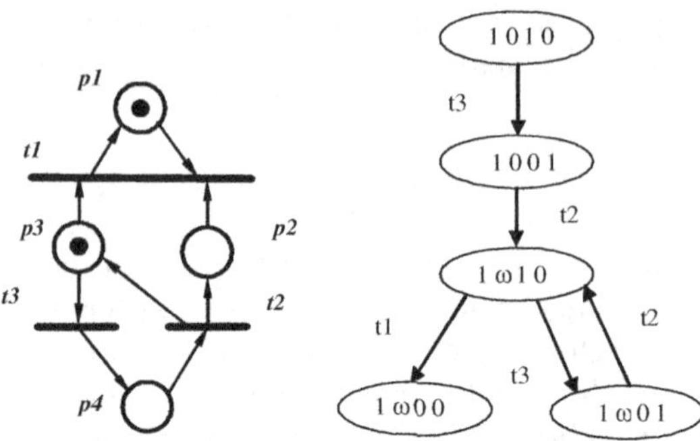

Fig. 2.8. A Petri net and its coverability graph[1]

Substitution: let $\pi \subset P$ be such set of places that $M_0(\pi) = \emptyset$, $\forall t \in T((^{\bullet}t \cap \pi \neq \emptyset) \Rightarrow (^{\bullet}t = \pi))$, $\forall t \in T((t^{\bullet} \cap \pi \neq \emptyset) \Rightarrow (\pi \subseteq t^{\bullet}))$, $\exists t \in T(t^{\bullet} \cap \pi \neq \emptyset)$. Then the net can be transformed by removing all transitions t_i such that $^{\bullet}t_i = \pi$, and replacing every transition t_j such that $\pi \subseteq t_j^{\bullet}$ by the set of transitions obtained from it by substitution to $t_j^{\bullet}$ instead of subsets π the sets $t_i^{\bullet}$ from those transitions t_i, where $^{\bullet}t_i = \pi$.

It is shown, that those rules preserve liveness and safeness of the nets [235, 249], and that they reduce every LSα-net to the net with single place and single transition [208, 243]. For further information on this subject, the reader is referred to [235, 243, 249]. More general results, describing reduction rules which reduce all and only structurally live and bounded free choice nets to the nets with one place and one transition, are presented in [55].

The **state equation** method [172, 183], the most known one among the **integer programming** methods based on incidence matrix of the Petri net, checks a necessary condition for the reachability of a marking M. The *state equation* for a Petri net can be written as follows:

$$M = M_0 + Cx, \tag{2.3}$$

where markings are written as the column vectors. If M is reachable from M_0 in the Petri net, then there exists a non-negative integer solution of (2.3), being the Parikh vector of a firing sequence σ such that $M_0 \sigma M$. The next example presents transformation of the initial marking of the net shown in Fig. 2.3 into the marking $\{p_1, p_2, p_3\}$ by the firing sequence $t_3 t_2$:

$$
\begin{bmatrix} 1 \\ 1 \\ 1 \\ 0 \end{bmatrix}
=
\begin{bmatrix} 1 \\ 0 \\ 1 \\ 0 \end{bmatrix}
+
\begin{bmatrix} 0 & 0 & 0 \\ -1 & 1 & 0 \\ -1 & 1 & -1 \\ 0 & -1 & 1 \end{bmatrix}
\begin{bmatrix} 0 \\ 1 \\ 1 \end{bmatrix}.
\tag{2.4}
$$

---

[1] Note that here the nodes of coverability graph are presented as vectors, unlike the nodes of reachability graphs in other figures in this book, presented as sets.

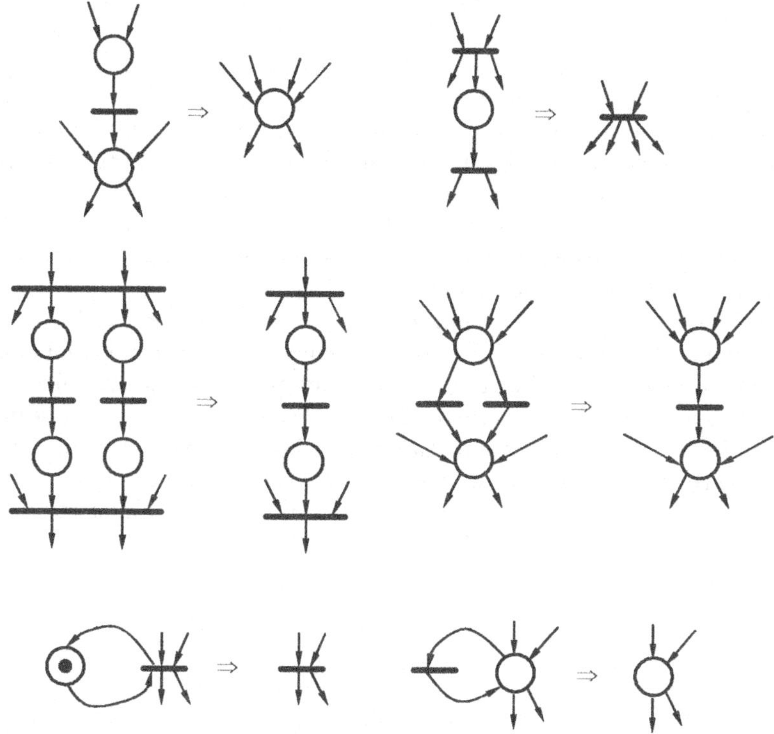

Fig. 2.9. Reduction transformations preserving liveness, safeness and boundedness

Some other methods using incidence matrix are mentioned in Section 5.1.

Compression techniques try to manage the state explosion problem by means of compact representation of state space. Kronecker algebra has been successfully used for representation of the state space of Petri nets [134]. But the most popular approach here is representing the sets of states of parallel discrete systems by their characteristic function. For such representation different versions of BDDs are usually used [34, 150, 151, 156, 181, 200]. BDD-based methods have proved to be efficient, however still there are many optimization problems, such as optimal ordering of variables, which affects the size of BDD greatly. In [168] representation of the state space of safe Petri nets by means of special monotonic Boolean function is proposed. It describes the state space unambiguously, and can be presented in more compact form than the characteristic function, but it is not clear, how to extract data from such representation efficiently. Encoding of places of Petri nets is used in some analysis approaches to minimize number of arguments of characteristic functions [45, 180]. Hierarchical decomposition of Petri nets helps to minimize representations of state space characteristic function [106, 167]. In [105, 138] obtaining the state space of sequent automata in form of ternary matrix is described.

Compression techniques can be successfully combined with the stubborn set method [217].

Another approach to avoid the state explosion problem is based on **partial order representations** of the behavior of a parallel system. It is known as the **unfolding method** [57, 58, 59, 74, 85, 86, 87, 165, 217]. This method represents state space not by a reachability graph, but by an unmarked Petri net (so-called *occurrence net*, or *unfolding*), where places correspond to the local states of a system, and the maximal sets of parallel places (where relation of parallelism is specially defined) correspond to the reachable global states. An occurrence net is acyclic by the construction (many places may correspond to the same local state), so for cyclic systems it is infinite and cannot be constructed as such. But its finite prefix (*finite unfolding*) can be constructed, capturing entire behavior of the system. In many cases (but not always) such prefix is much smaller than the reachability graph. Some properties can be checked easily from an unfolding, such as reachability of a state where a certain transition is enabled. Usually unfolding techniques are used to represent behavior of finite-state systems; however, in [1] an unfolding algorithm for infinite-state systems (unbounded Petri nets) is described.

3. Reduced Reachability Graphs

3.1 Review of Known Methods

Content of this section relates to the parallel discrete systems in general sense, unless otherwise stated.

3.1.1 Persistent Set Methods

The methods based on calculation of a subset of successors from a global state are known as "persistent set" methods. Informally, a *persistent set* is a subset T_P of transitions such that no transition firings outside T_P affect T_P. The formal definition is given below, but preliminarily we need the notion of independent transitions [77, 226].

Definition 3.1. *Transitions t_1, t_2 are independent in a global state M, if the following two conditions hold:*

1. *if t_1 (t_2) is enabled in M and Mt_1M' (Mt_2M'), then t_2 (t_1) is enabled in M, if and only if t_2 (t_1) is enabled in M' (independent transitions can neither disable nor enable each other);*
2. *if t_1 and t_2 are enabled in M, then there is a unique state M' such that Mt_1t_2M' and Mt_2t_1M' (commutativity of enabled independent transitions).*

Transitions t_1 and t_2 are globally independent, if they are independent in all reachable global states.

If the conditions do not hold, the transitions are said to be dependent (in a state or globally, correspondingly).

For the classical Petri nets and, generally, for all systems for which the diamond rule holds, the second condition is redundant. The diamond rule is a well-known property of Petri nets (see, for example, [22, 43]), which can be formulated as follows.

Diamond Rule: *If Mt_1t_2M' and Mt_2t_1M'', then $M' = M''$.*

A. Karatkevich: Dynamic Analysis of Petri Net-Based Discrete Sys., LNCIS 356, pp. 27–62, 2007.
springerlink.com

This property does not hold for many parallel models, such as some kinds of interpreted Petri nets.

The persistent set methods are often explained in terms of *Mazurkiewicz traces* [163]: a *trace* is a set of firing sequences, which can be obtained from each other by successively permuting adjacent independent transitions. One can say, that the persistent set methods tend to represent a trace with only one of firing sequences belonging to it (one of its *linearizations*) [77].

Definition 3.2. *[77, 78] A set T_P of transitions enabled in a state M is persistent in M if and only if, for all sequences in the full reachability graph $M_1 t_1 M_2 t_2 M_3...t_{n-1} M_n t_n M_{n+1}$ starting from M ($M = M_1$) and including only transitions $t_i \notin T_P$ ($1 \leq i \leq n$), t_n is independent in M_n of all transitions in T_P.*

A *selective search* operates as a classical state space search except that, for each state M reached during the search, it computes a subset of the set of transitions that are enabled in M, and explores only the transitions in this subset, the other enabled transitions being not explored. Let a *persistent-set selective search* be a selective search of the reachability graph which, for each state M that it reaches, selects a nonempty set of enabled transitions T_P that is persistent in M. Then the next theorem holds [77, 226].

Theorem 3.3. *Let M be a state reached in a persistent-set selective search, and let M_d be a deadlock. If M_d is reachable from M, then M_d will also be reached by the persistent-set selective search[1].*

The **stubborn set method** developed by A. Valmari [212, 215] is considered as the most elaborate technique in the family of persistent set methods[2]. Its main notion is the *stubborn set*, defined for classical Petri nets as follows.

Definition 3.4. *A set T_S of the transitions of a Petri net at marking M is a stubborn set, if (1) every disabled transition in T_S has an empty input place p such that all transitions in $\bullet p$ are in T_S; (2) no enabled transition in T_S has a common input place with any transition (including disabled ones) outside T_S; and (3) T_S contains an enabled transition.*

Strictly speaking, this is one of possible definitions of a *strong stubborn set*; for the alternative definitions see [217]. Below, whenever the stubborn sets are mentioned, we talk about the strong stubborn sets, if not stated otherwise. There also exist the notion and theory of *weak stubborn sets* [212, 215, 216, 217], which provide better possibilities for reduction of state spaces, but are more difficult to construct. One of possible definitions of weak stubborn sets is the following.

---

[1] In [77, 226] Theorem 3.3 is proved for the *labelled formal concurrent systems* (LFCS). This model is less general than the parallel discrete systems defined in subsection 2.1.1. But it is easy to show that the Theorem holds for the general model; it is enough to directly apply the corresponding proofs (of Lemma 4.2 and Theorem 4.3 from [77]) to it.

[2] Historically, the theory of stubborn sets arose before the theory of persistent sets.

Definition 3.5. *A set T_{wS} of the transitions of a Petri net at marking M is a weak stubborn set, if (1) every disabled transition in T_{wS} has an empty input place p such that all transitions in ${}^\bullet p$ are in T_{wS}; (2) for every enabled transition t in T_{wS} all the transitions with which t shares input places **or** all the transitions for which output places are the input places of t (including disabled ones) are in T_{wS}; and (3) T_{wS} contains an enabled transition, which has no common input place with any transition (including disabled ones) outside T_{wS}.*

The next properties hold for (strong) stubborn sets [217].

Lemma 3.6. *if $t \in T_S$, $t_1, ..., t_n \notin T_S$, $M_0 t_1 t_2 ... t_n M_n$, and $M_n t M_n'$, then there is M_0' such that $M_0 t M_0'$ and $M_0' t_1 t_2 ... t_n M_n'$.*

Lemma 3.7. *If transition $t_k \in T_S$ is enabled in M_0, $t_1, ..., t_n \notin T_S$ and $M_0 t_1 t_2 ... t_n M_n$, then t_k is enabled in M_n.*

Lemma 3.8. *If T_S is a stubborn set at M, $t_1, ..., t_n \notin T_S$, $t_k \in T_S$ is enabled in M, and firing sequence $t_1 t_2 ... t_n$ is enabled in M, then there exists M' such that $M t_k M'$ and $t_1 t_2 ... t_n$ is enabled in M'.*

The basic stubborn set method builds a reduced reachability graph (RRG) in the following way: for every considered state T_S is calculated, and only firing of enabled transitions belonging to T_S is simulated. Below RRG means a reduced reachability graph that has been constructed using the basic stubborn set method, unless otherwise stated.

The next statement describes the fundamental property of the stubborn set method [217].

Theorem 3.9. *RRG contains all deadlocks of the system that are reachable from the initial states. Furthermore, all deadlocks of the RRG are deadlocks of the system.*

Relation between stubborn sets and persistent sets is described by the next theorem [77].

Theorem 3.10. *The set of all enabled transitions of a stubborn set is a persistent set.*

Relation between weak stubborn sets and persistent sets is discussed in Section 3.3.

So, the stubborn set method allows to detect all reachable deadlocks of a Petri net. It also allows to detect at least one infinite firing sequence, if the net has any (this property had been mentioned for the first time in [210]).

The method is unable - directly and for a net without restrictions - to analyze most of other important net properties because of the so-called *ignoring problem* (the method may generate infinite firing sequence, infinitely ignoring the enabled transitions not belonging to the stubborn sets [215, 217]). However, there are some additional possibilities of the basic stubborn set method, especially when certain restrictions are imposed on the net. It will be shown in the next section of this chapter. Formally, ignoring can be defined as follows.

Definition 3.11. *Ignoring of transition t occurs in the RRG, if there is a state M in it such that t is enabled in M and there is no path in the RRG corresponding to firing sequence σ such that $M\sigma M'$, $t \in T_S$ for M'.*

The methods solving the ignoring problem have also been developed (see, for example, [212]). The main idea is to "force" the ignored transitions firing, to guarantee that all the "interesting" firing sequences have been investigated.

The question of optimal construction of stubborn sets is of great practical importance, because size of RRG may considerably depend on it. However this question is beyond the scope of the book.

In Fig. 3.1 an example of the applied method is shown. For comparison, in Fig. 3.2, the full reachability graph for the same net is shown (in the nodes of reachability graphs the numbers of marked places are listed).

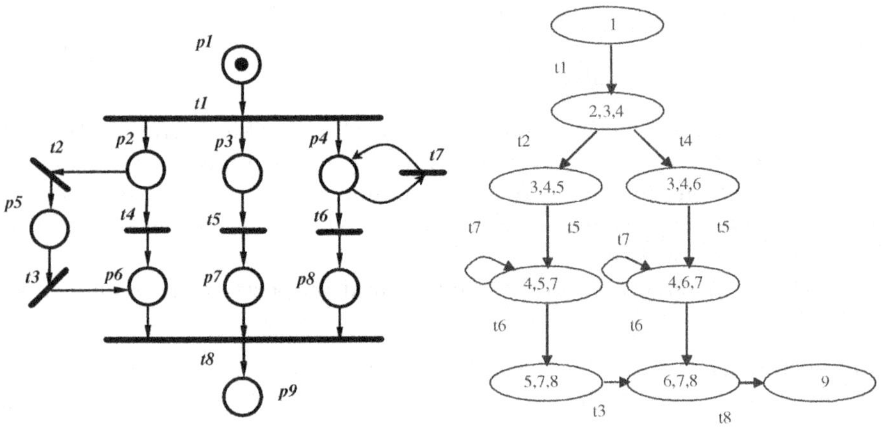

Fig. 3.1. A Petri net and RRG for it

It is easy to see that both graphs contain the same set of deadlocks (one-element in this case), but the first one is much smaller and simpler. It is difficult, however, to evaluate in general case the difference between RRG and the corresponding full reachability graph, because it depends greatly on the net structure and selected stubborn sets.

Another similar approach to analysis of Petri nets and, generally, of parallel discrete systems, is based on the idea of **decomposition** [19,175,251]. The task of net analysis is reduced to the task of analysis of the *blocks* of its decomposition, which may be considerably smaller than the net itself [249]. This approach can simplify analysis of large nets.

Definition 3.12. *A subnet $\Sigma_j = (P_j, T_j, F_j)$ of a Petri net $\Sigma = (P,T,F)$ is a block, if and only if $\forall t_i \in T \setminus T_j : t_i^{\bullet} \cap P_j \subseteq P_j^{in}, {}^{\bullet}t_i \cap P_j \subseteq P_j^{out}$, where $P_j^{in} = {}^{\bullet}T_j \setminus T_j^{\bullet}$ is the set of input places, $P_j^{out} = T_j^{\bullet} \setminus {}^{\bullet}T_j$ is the set of output places.*

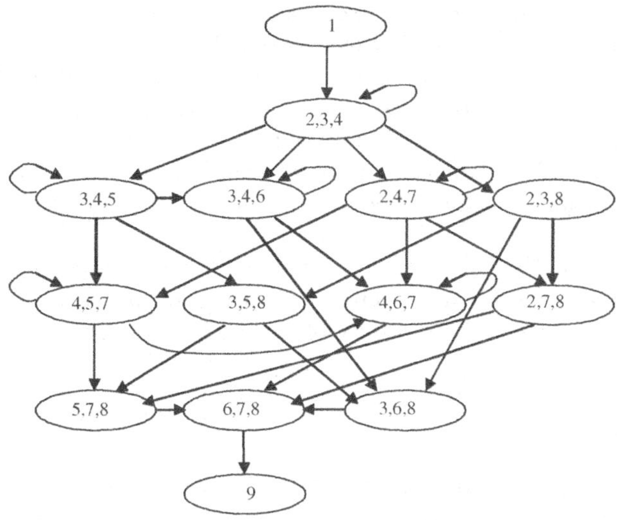

Fig. 3.2. Full reachability graph for the net from Fig. 3.1

Informally, a subnet is a block, if any transition outside it may be incident only to the input or output places of the subnet, and in the first case they are the output places of the transition, in the second case they are its input places.

Definition 3.13. *Two transitions are in the relation of alternative joint if their input or output sets of places intersect.*

Lemma 3.14. *Transitive closure of the relation of alternative joint of the set of transitions of a Petri net specifies its decomposition into minimal blocks.*

The Zakrevskij's method of analysis (in particular, deadlock detection) of Petri nets, using block decomposition, is described in details in Chapter 4 (and also in [249, 251]).

The **ample set method** proposed by Peled [31, 44, 45, 182] is based on similar ideas as the stubborn set method, but has been designed to check different properties (which can be expressed in the temporal logic LTL with the next-time operator omitted) and takes into account the output signals of a system. An *ample set* is defined below.

Definition 3.15. *A set ample(M) of the transitions of a parallel system for global state M is an ample set, if (0) ample$(M) = \emptyset$ iff enabled$(M) = \emptyset$; (1) in the full reachability graph, on any path starting from state M, a transition dependent on a transition from ample(M) cannot appear before some transition from ample(M) is executed; (2) if M is not fully expanded, every transition $t \in$ ample(M) has to be invisible; and (3) a cycle (in the reachability graph) is not allowed if it contains a state in which some transition t is enabled, but it is never included in ample(M') for any state M' on the cycle.*

A transition is *invisible* when its execution does not change the value of the output variables of the system. For details see [44, 45].

It is evident, that unlike the definition of stubborn set, definition of ample set is not constructive - condition (1) seems to require knowledge about the full reachability graph. To avoid constructing the full graph, the algorithms have been developed producing sets of transitions for which condition (1) is guaranteed [2, 44, 45]. Another difference is that condition (3) prevents occurring an ignoring.

The ample set method is suitable, in particular, for solving safety properties [182] or coverability problem [2]. The method has been primarily implemented in the SPIN model-checking tool developed by G. Holzmann and D. Peled [44, 45, 89, 90].

3.1.2   Other Methods

The **sleep set method** proposed by P. Godefroid [76, 77, 226] is based on the following idea: if firing of two or more independent transitions in any order leads to the same state, then it is enough to simulate one of these orders. The strategy is to compute, in each state M encountered, a set of transitions which are enabled but will *not* be explored from M (a *sleep set*). The sleep set for a given marking is calculated from the sleep sets of its predecessor markings, the independence relation, and the order in which enabled elements are explored in the predecessor markings. The method has been initially developed for deadlock detection.

This method, in its pure form, reduces only arcs in reachability graph and preserves all its nodes, hence it is not of much help. But it is compatible with the stubborn set method, and the combination leads to better results than each of the methods alone [77, 219].

The **symmetry method** [45, 93, 145, 187, 220] is applicable to the systems containing several identical (or nearly identical) components, called *symmetric* (such situation is rather common). The main idea is to construct, instead of the full state space, *a condensed state space* in which all states which differ only by permutation of symmetric components are presented as one state. Such condensed state space may be orders of magnitude smaller than the full state space, but the same dynamic properties can be directly analyzed from it [145].

The symmetries may be calculated automatically, but it is a time-consuming task. On the other hand, the analyst (system designer) can usually select symmetric components in the model [187].

One interesting approach, offering unfortunately limited applicability, is the **maximal concurrent simulation**. Its rule is: "always choose a maximal set of independent transitions to be executed next" [99]. The approach is not sufficiently expressive in general case, because it may lead to removing tokens from input places of a transition without further simulation of this transition firing. This is not the case for free choice systems, so maximal concurrent simulation is completely applicable for EFC-nets. In Fig. 3.3 the reduced reachability

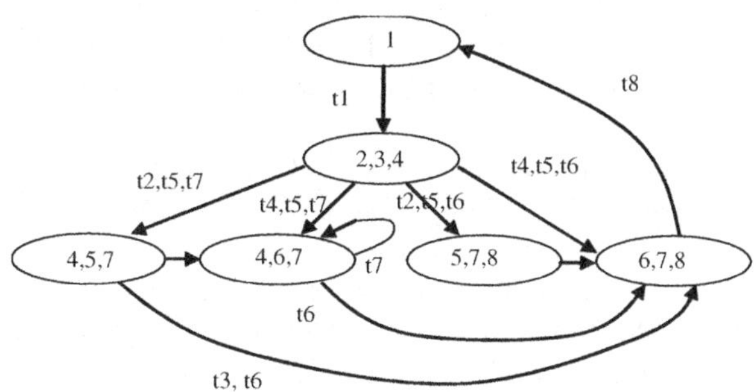

Fig. 3.3. RRG for the net of Fig. 3.1, constructed with maximal concurrent simulation

graph for the net from Fig. 3.1 built using maximal concurrent simulation approach is shown.

In [97] a more refined method of optimal simulation is described, based partially on maximal concurrent simulation approach and allowing to decide a number of interesting system properties such as deadlock-freeness and liveness.

3.2 A Generalization of Stubborn Set Method

Notion of independent transitions is important for the persistent set methods. But, as it can be seen from Definition 3.4, it is also essential to distinguish between the cases when one of dependent transitions can *disable* another one and when one of them can *enable* another. For the systems for which the diamond rule does not hold it is important also to consider the kind of dependency which breaks the second condition of Definition 3.1.

The definition of stubborn sets can be re-formulated using different kinds of dependency between transitions. Below we formulate the definition for generalized parallel systems.

Definition 3.16. *A set T_S of the transitions of a parallel discrete system at global state M is a stubborn set, if (1) for every disabled transition in T_S all the transitions that can enable[3] it are in T_S; (2) for every enabled transition in T_S all the transitions it can disable, all the transitions which can disable it, and all the transitions which are not commutative in respect to it are in T_S; and (3) T_S contains an enabled transition.*

The next theorems constitute theoretical base of the stubborn set method for the generalized parallel systems.

Theorem 3.17. *Let T_S be a stubborn set at a global state M according to Definition 3.16. Then the set T_P of all enabled transitions in T_S is a persistent set.*

---

[3] Here by "t can enable (disable) t'" we mean "exists a global state, in which firing of t enables (disables) t'".

Proof. According to the definition of persistent sets (Definition 3.2), it is enough to prove that for any sequence in the full reachability graph $Mt_1t_2...t_{n-1}M_nt_n$ such that $t_i \notin T_P$ $(1 \le i \le n)$ transition t_n is independent in M_n with respect to all transitions in T_P .

The proof proceeds by induction on n. Let $n = 1$ (then $M_n = M$). Every enabled transition outside T_S is independent of any enabled transition in T_S by definition (condition (2) of Definition 3.16), and for $n = 1$ the theorem holds.

Now, assume that the theorem holds for every sequence of length $(n-1)$ and let us prove that it holds for a sequence of length n. Suppose transition t_n is dependent on a transition $t \in T_P$ in M_n. It follows from the inductive hypothesis, that $t \in enabled(M_n)$. Then the next variants are possible: (1) t_n can disable t; (2) t_n can be disabled by t; (3) t_n and t are not commutative. In all three cases, according to Definition 3.16, $t_n \in T_S$. But $t_n \notin T_P$, hence $t_n \notin enabled(M)$, and then there is transition t_i $(0 \le i < n)$ which enables t_n. According to Definition 3.16, $t_i \in T_S$. If $t_i \in enabled(M)$, then there is contradiction with assumption that $t_i \notin T_P$. If $t_i \notin enabled(M)$, then there is transition t_j $(1 \le j < i)$ enabling t_i, and the same reasoning leads to the contradiction.

Theorem 3.18. *Reduced reachability graph of a parallel system, created in such way that in every considered global state only firing of enabled transitions belonging to a set T_S being stubborn in sense of Definition 3.16 is simulated, contains all deadlocks of the system that are reachable from the initial states. Furthermore, all deadlocks of the RRG are deadlocks of the system.*

Proof of the first part follows directly from Theorem 3.17 and Theorem 3.3. The second part follows from the third condition of Definition 3.16.

Applying Definition 3.16 to the Petri nets, we can note that it is stronger than Definition 3.4; the first condition of Definition 3.16 in terms of Petri nets would look as follows: *for every disabled transition in T_S and its every input place p all transitions in $^\bullet p$ are in T_S.* Practically it means, that the stubborn sets for Petri nets, according to Definition 3.16 will be the same, as according to Definition 3.4, when the stubborn sets consist of the enabled transitions only (as in the case of EFC-nets), and can be remarkably bigger otherwise. Definition 3.16 can be, however, re-formulated in such a way, that applied to the Petri nets it would be equivalent to Definition 3.4. In this variant (let us call it **Definition 3.16a**) condition (1) looks as follows: *for every disabled transition in T_S and an unsatisfied necessary condition of its enabling, all the transitions firing of which can satisfy this condition are in T_S;* the rest of conditions remain as in Definition 3.16. It is easy to see, that Theorem 3.17 and Theorem 3.18 hold for Definition 3.16a.

3.3 Weak Persistent Sets

Persistent set methods in many cases successfully avoid interleaving. Sometimes, however, they fail to do that. Consider, for example, the following net (Fig. 3.4).

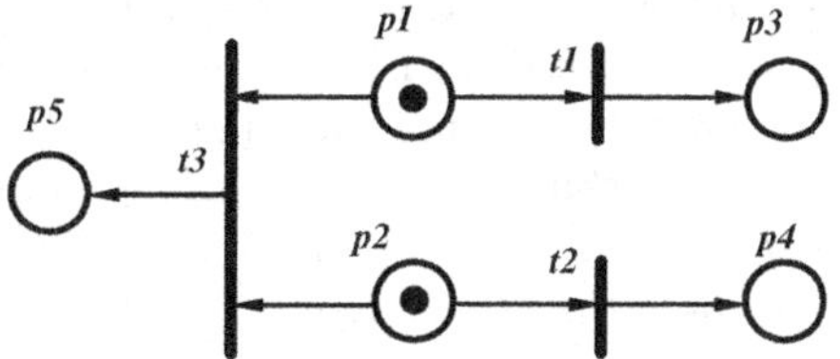

Fig. 3.4. A Petri net

Here, according to Definition 3.2, the only persistent set is $\{t_1, t_2, t_3\}$, but transitions t_1 and t_2 are independent (see Definition 3.1), and execution of both of them from the marking shown in Fig. 3.4 leads to interleaving, which the persistent set approach tries to avoid.

In the example under discussion, any element from the pair of independent transitions could be dropped from the persistent set, and the same deadlock will be reached in such reduced search. However it is not always the case.

Below we present the definition of *weak persistent set*, having potential of better reduction than the classical persistent sets (compare with Definition 3.2)[4].

Definition 3.19. *A set T_{wP} of transitions enabled in a marking M of a Petri net[5] is a weak persistent set, if (1) for any firing sequence in the full reachability graph $Mt_1t_2...t_n$ such that $t_i \notin T_{wP}$ $(1 \leq i \leq n)$ there is a transition $t \in T_{wP}$ such that any transition t_i is independent of t and all enabled transitions dependent on t belong to T_{wP}, and (2) in the full reachability graph there is no sequence $Mt'_1t'_2...t'_m M_m t'_{m+1}...t'_k M_k$ such that $t'_j \notin T_{wP}$ $(1 \leq j \leq k)$ and a transition $t' \in T_{wP}$ at M is disabled in M_m and enabled in M_k[6], and (3) no transition in T_{wP} has an input place p which is at the same time input and output place of a transition outside T_{wP}.*

Lemma 3.20. *Every persistent set is also a weak persistent set, but not vice versa.*

Proof. $\Rightarrow$ Follows from the definitions.
$\Rightarrow$ See Fig. 3.4: $\{t_1, t_3\}$ is a weak persistent set, but not a (Godefroid's) persistent set.

Lemma 3.21. *Let M be a marking, $Mt_1M_1t_2M_2...t_nM_n$ be a sequence in the reachability graph, t is a transition enabled in all markings $M, M_1, ..., M_n$, and there is no such transition t_i in the sequence that $\bullet t_i \cap t_i^\bullet \cap \bullet t \neq \emptyset$. Then the sequence $tt_1t_2...t_n$ is enabled in M and leads to the same marking M' as the sequence $t_1t_2...t_nt$.*

---

[4] In [77] (Note 4.31) Godefroid writes: "...following the idea of Valmari, "weak" versions of our notions of persistent set ... can easily be defined". In fact that is what Section 3.3 about.

[5] Content of this section relates only to the ordinary Petri nets.

[6] Informally, the second condition prevents $t' \in T_{wP}$ from first being disabled and later enabled again by transitions outside T_{wP}.

Proof. The effect of the firing of a transition is an addition of an integer valued vector to the marking. Vector addition is commutative, so if two sequences of transition firing are enabled in the same marking and differ only in the order of transitions, the resulting marking will be the same. So, it is enough to prove that the firing sequence $tt_1t_2...t_n$ is enabled in M.

The proof proceeds by induction on n. Let $n = 1$. Suppose that t disables t_1. Then t and t_1 share an input place p, which has 1 token in M. But as far as t_1 does not disable t, $p \in t_1^\bullet$, which contradicts the assumption $^\bullet t_i \cap t_i^\bullet \cap {}^\bullet t \neq \emptyset$. So, t and t_1 do not disable each other, and for $n = 1$ the lemma holds.

Now, assume that the lemma holds for every sequence of length $(n - 1) > 0$ and let us prove that it holds for a sequence of length n. It follows from the inductive hypothesis, that sequence $tt_1t_2...t_{n-1}$ is enabled in M. It leads to the same marking M'' as $t_1t_2...t_{n-1}t$. Firing of t from M_{n-1} does not disable t_n, which follows from the same reasoning as at the first step of induction. Hence the sequence $Mt_1t_2...t_{n-1}tM''t_nM'$ exists. From the above it follows also that the sequence $Mtt_1t_2...t_{n-1}M''t_nM'$ exists, so the lemma holds for n. This together with the inductive hypothesis proves the lemma.

Let a *weak-persistent-set selective search* be a selective search of the reachability graph which, for each marking M that it reaches, selects a nonempty set of enabled transitions T_{wP} that is a weak persistent set at M.

Theorem 3.22. *Let M be a marking reached in a weak-persistent-set selective search, and let M_d be a deadlock. If M_d is reachable from M, then M_d will also be reached by the weak-persistent-set selective search.*

Proof. Let $Mt_1M_1t_2M_2...t_nM_d$ be a sequence in the full reachability graph. The proof proceeds by induction on n. For $n = 1$ if $t_1 \in T_{wP}$, the result is immediate; supposing $t_1 \notin T_{wP}$, we obtain that there exists a transition t enabled in M and independent of t_1, then t is enabled in M_d, which contradicts the assumption that M_d is a deadlock.

Now, assume the theorem holds for every sequence of length $(n-1) > 0$ and let us prove that it holds for a sequence of length n. If no transition in the sequence belongs to T_{wP}, then there exists a transition t enabled in M and still enabled in M_d, which contradicts the assumption that M_d is a deadlock. So, there is at least one transition in the sequence, which belongs to T_{wP}. Let t_i be the first such transition. Then t_i is enabled in all markings between M and M_{i-1} (condition (2) from Definition 3.19) and in the sequence $t_1...t_{i-1}$ there is no transition t_k such that $^\bullet t_k \cap t_k^\bullet \cap {}^\bullet t_i \neq \emptyset$ (condition (3) from Definition 3.19). From Lemma 3.21 the sequence $Mt_iM't_1...t_{i-1}M_it_{i+1}...t_n$ exists in the full reachability graph. But $t_i \in T_{wP}$, hence M' will be reached by the weak-persistent-set selective search. As far as a firing sequence of length $n - 1$ from M' to M_d exists in the full reachability graph, it follows from the induction assumption that M_d will be reached by the weak-persistent-set selective search. This together with the inductive hypothesis proves the theorem.

Relation between weak persistent sets and weak stubborn sets (for safe Petri nets) is described by the next theorem.

Theorem 3.23. *Let T_{wS} be a weak stubborn set of a Petri net at marking M (according to Definition 3.5). Then the set T_e of all enabled transitions of T_{wS} is a weak persistent set.*

Proof. We have to prove, that for any sequence $Mt_1M_1...t_nM_nt_{n+1}$, such that $t_i \notin T_e$, 3 conditions of Definition 3.19 hold.

1. For the first condition it is enough to prove, that transition $t_k \in T_{wS}$ such that t_k is enabled in M and has no common input place with any transition outside T_{wS}, is independent of any transition t_i in the marking M_{i-1} in the sequence mentioned above. Suppose there are transitions in the sequence, dependent on t_k, and let t_m be the first such transition. Then t_m has a common input place with t_k (as far as both t_m and t_k are enabled in M_{m-1}, there is no other possibility), $t_m \in T_{wS}$, $t_m \notin enabled(M)$ (otherwise $t_m \in T_e$), and for any input place p of t_m empty at M there is a transition t_i ($1 \leq i < m$) such that $p \in t_i^\bullet$. At least one of such transitions $t_i \in T_{wS}$. If t_i is enabled in M, then $t_i \in T_e$, and there is a contradiction. If t_i is disabled in M, the same reasoning can be applied to it as to t_m above, until we'll come to the contradiction, considering transition t_1.
2. From Definition 3.5, for any enabled transition t in T_{wS} (which by construction belongs to T_e) every transition, which can enable it, **or** every transition, which can disable it, is in T_{wS}. Supposing that sequence of the transitions outside T_e can disable, and later enable t, we come to the conclusion, that at least one transition in this sequence is in T_{wS} and is disabled in M. And from the reasoning given in item 1 we see, that it leads to a contradiction.
3. Condition 3 is satisfied by Definition 3.5.

3.4 On Combining the Persistent Set Approach and Concurrent Simulation

3.4.1 Concurrent Simulation and Persistent Sets

Is it possible to unite in one method, advantages of the persistent set approach and of concurrent simulation? Direct combination of these two approaches seems to be impossible, because the main ideas are almost opposite: generally, if two non-conflicting transitions are simultaneously enabled, the first approach supposes that their firing will be simulated one-by-one, and the second - simultaneously.

But, as it was mentioned in Section 3.1, the maximal concurrent simulation approach is not sufficiently expressive; for example, it is easy to show, that, being directly applied, it may fail to detect some deadlocks. (See Fig. 3.5; here maximal concurrent simulation investigates the only step sequence $\{t_1, t_2\}\{t_4\}$, and the deadlock with token in p_5 will remain undetected). So, maybe the concurrent simulation approach can be improved by the idea of persistent sets? We claim that the answer is positive. In [109] a very particular case of this possibility is presented; here we present a generalization of the idea described there.

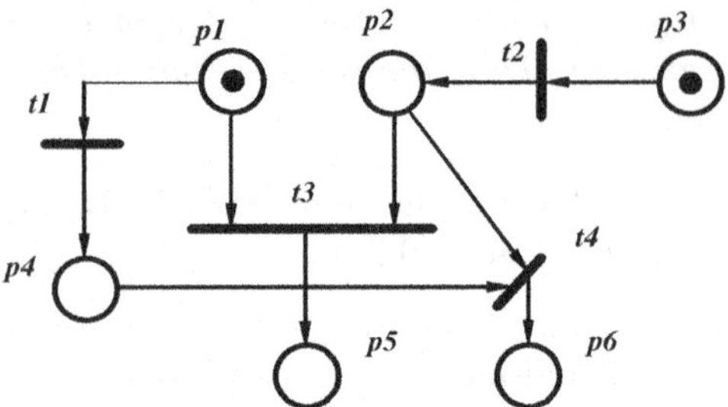

Fig. 3.5. A Petri net (taken from [98])

Theorem 3.24. *Let $\Sigma = (P, T, F)$ be a safe Petri net, let $U = \{T_{P1}, T_{P2}, ...T_{Pn}\}$ be a set of non-intersecting persistent sets at M_0. Execute from M_0 every step $\Delta_k = \{t_1^k, t_2^k, ..., t_n^k\}$ such that $t_i^k \in T_{Pi}$. Repeat the operation for every newly obtained marking. Every deadlock reachable from M_0 in Σ will be reached by such search.*

Proof. Let M_d be a reachable deadlock. Then exists sequence $M_0 \sigma M_d$. As far as no transition is enabled in M_d and no transition outside a persistent set can disable a transition in the persistent set, for every $T_{Pi} \in U$ there is a transition $t_i \in \sigma$, such that all previous transitions in σ are independent of it; all such transitions (belonging to different persistent sets) are mutually independent. From Lemma 3.21, every such transition can be moved to the beginning of the sequence, and the sequence will remain enabled and leading to the same marking. Then there exists a sequence $M_0 t_1' t_2' ... t_n' \sigma' M_d$ being an interleaving of σ, such that there is a step $\Delta_1 = \{t_1', t_2', ...t_n'\}$, which will be executed from M_0. So, $M_0 \Delta_1 M_1 \sigma' M_d$. If $|\sigma'| = 0$, then $M_1 = M_d$ and the theorem is proved; otherwise the same reasoning, as was applied above to M_0 and M_d, can be repeated for the markings M_1 and M_d. As far as $|\sigma'| < |\sigma|$, M_d will be reached in a finite number of steps.

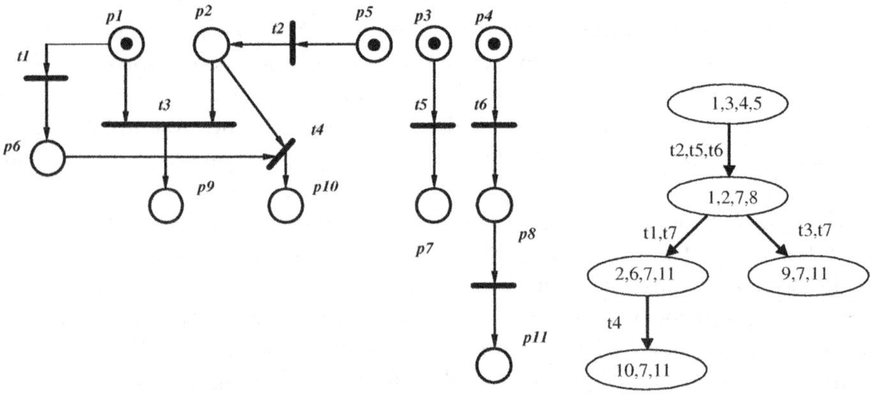

Fig. 3.6. A net (taken from [98]) and reachability graph generated by PSS

Below we will refer to the combined method of simulation described in Theorem 3.24 as *parallel selective search* (PSS). An example of applying of PSS is shown in Fig. 3.6.

Consider the net shown in Fig. 3.4. PSS does not allow in this case to simulate transitions firing in parallel, because there is only one persistent set. However it seems to be intuitively clear, that firing of transitions t_1 and t_2 (belonging to the same strong stubborn set) can be simulated in parallel. They are independent, and independency seems to be a necessary condition for parallel simulation. But this condition is not sufficient, as illustrated by Fig. 3.5: transitions t_1 and t_2 are independent, but their parallel simulation makes impossible detection of one of the deadlocks.

The next theorem provides a theoretical possibility to capture such situations.

Theorem 3.25. *Let $\Sigma = (P, T, F)$ be a safe Petri net, let $M_0 \Delta_1 M_1 \Delta_2 ... \Delta_n M_d$ be a step sequence, and let M_d be a deadlock. Let $T_p \subset enabled(M_0)$ be a set which is persistent at M_1 (not necessary at M_0), and all transitions in T_P are independent of all transitions in Δ_1. Then exists transition $t \in T_p$ such that $M_0 \Delta M$, where $\Delta = \Delta_1 \cup \{t\}$, and firing sequence σ such that $M \sigma M_d$ and $|\sigma| < \sum_{i=2}^{i=n} |\Delta_n|$.*

Proof. As far as no transition is enabled in M_d and no transition outside a persistent set can disable a transition in the persistent set, exists transition $t' \in T_P$ such that $t' \in \Delta_i$, where $2 \leq i \leq n$. Let $t = t'$. From Definition 3.2, without lost of generality we can suppose, that t is enabled in all markings $M_0...M_{i-i}$ and every transition $t'' \in \Delta_j$ such that $1 \leq j < i$ is independent of t. Let σ' be a linearization of $\Delta_1 \Delta_2 ... \Delta_n$. From Lemma 3.21, t can be moved to the beginning of the sequence, and the sequence (let it be σ''; $|\sigma''| = |\sigma'|$) will remain enabled and leading to the same marking. By the construction, t is the first transition in σ'', and the next $|\Delta_1|$ transitions (a linearization of Δ_1) are independent of t and can be executed in parallel from M_0. Then $M_0 \Delta M \sigma M_d$, where σ is σ'' without first $|\Delta|$ transitions, and $|\sigma| = \sum_{i=2}^{i=n} |\Delta_n| - 1$.

It is not clear however, how to compute a set T_P from Theorem 3.25 without previous simulation of Δ_1; and without a method of such calculation practical application of Theorem 3.25 for improvement of PSS is doubtful.

3.4.2 Comparison with the Janicki-Koutny's Method

It is interesting to compare PSS with the method of concurrent simulation by R. Janicki and M. Koutny, described in [96,98]. Janicki and Koutny claim [98], that their method provide **the** optimal simulation (OPT), because:

1. Some behavioral properties are common to full simulation and optimal simulation by their method; for example, they generate the same set of deadlocks.

2. OPT involves a minimum set of step sequences; each proper subset of OPT is less expressive and may not be used, e.g., to verify the deadlock-freeness.
3. OPT uses the shortest step sequences. Each deadlock will be generated by following shortest possible step sequence leading to it.

Let us compare the reachability graphs generated by OPT and PSS. The corresponding graphs are shown in Fig. 3.6-3.9.

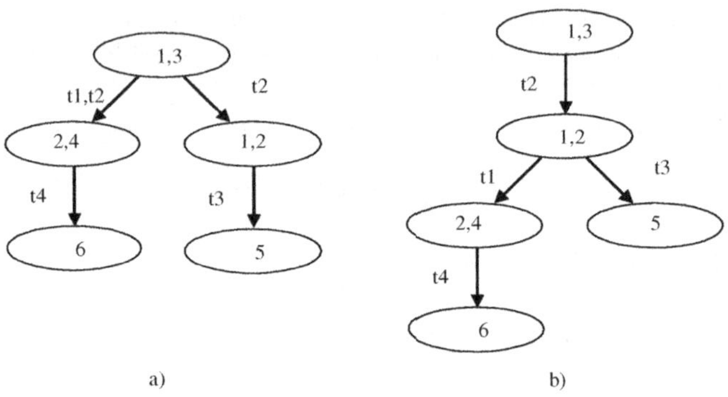

Fig. 3.7. Reachability graphs generated by OPT (a) and by PSS (b) for the net from Fig. 3.5

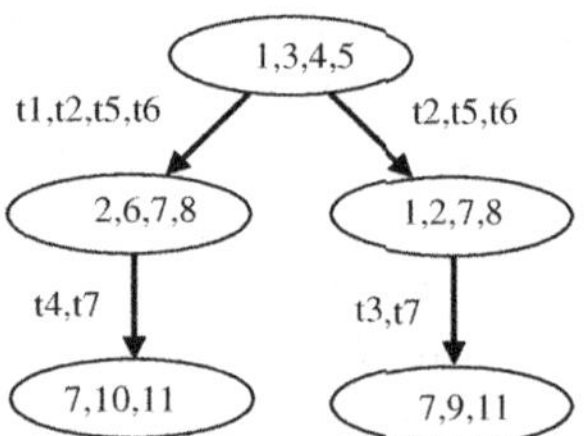

Fig. 3.8. Reachability graphs generated by OPT for the net from Fig. 3.6

The essential differences between PSS and OPT are:

1. OPT generates shorter firing sequences, but PSS simulates in some cases less transition firings.
2. OPT is defined only for the state machine decomposable (SMD) nets, PSS works for any safe Petri net (probably for any ordinary Petri net).
3. In general case in a reachability graph generated by OPT there may exist different nodes for the same marking. PSS constructs the graph in which no more than one node corresponds to a marking.
4. In some cases PSS (not taking into account Theorem 3.25) fails to avoid an interleaving which is avoided by OPT (Fig. 3.9)

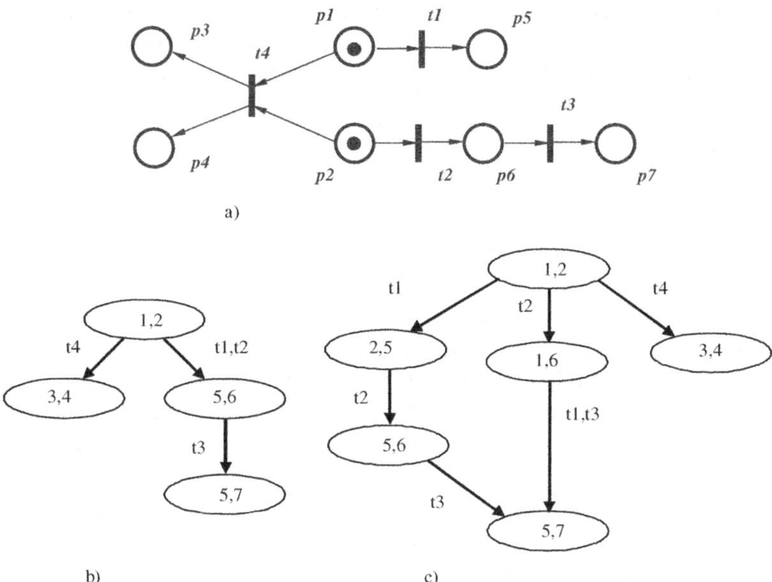

Fig. 3.9. A Petri net from [98] (a), its reduced reachability graphs generated by OPT (b) and PSS (c)

3.4.3 Conflicts, State Explosion and Decomposition

Concurrent simulation cannot avoid state explosion if the Petri net contains lots of conflicts, because such simulation would require checking of all mutual combinations of alternative variants in all parallel branches. That is the same problem as with unfolding of such nets [217]. Consider a net with r parallel branches, each consisting of n alternative branches and having length q. The full reachability graph would consist of $(nq)^r$ nodes, RRG with maximal concurrent simulation - of $n^r q$ nodes, and RRG created by stubborn set method - of nqr nodes. It means that comparative effectiveness of those methods does not depend on the length of branches but does depend on the extent of conflicts.

This problem seems to be in certain respect similar to the problem of avoiding interleaving. Different ordering of firing of independent transitions leads to the same result, so why consider all mutual combinations? This is the background of persistent set approach. On the other hand, sometimes the final result (i.e. a deadlock) does not depend on alternative choices in the parallel branches of a system. In terms of blocks (Definition 3.12) it can be formulated as follows: the choices made inside the blocks affect the behavior of the rest of the system only through terminal markings of the blocks. Why then should we consider all mutual combinations of these choices?

The net of Fig. 2.1 has 3 parallel branches, each of which consists of 2 alternative branches, starting and ending in the same place. Such *two-pole blocks* are typical for structures of control and computational algorithms [249]. The parallel simulation approach, as it has been described so far, would check 2x3=6 variants of execution, but there is only one reachable deadlock, and intuitively it seems to be clear, that there is no sense to check all these variants.

The idea of blocks and block decomposition together with the idea of concurrent simulation allows to develop analysis method excluding such checking. It is described in Chapter 4.

3.5 Analysis of Special Classes of Petri Nets

Usually the parallel algorithms start from a sequential process (or single step), which branches later into several parallel processes. Therefore some languages for parallel control algorithms (SFC, PRALU etc.) require single initial step. The same is true for parallel computational algorithms. The underlying Petri nets of such algorithms usually have single-token initial markings. Such nets are called below the *s-nets*.

On the other hand, free choice nets and EFC-nets [55,82] constitute the especially convenient classes of Petri nets for describing structure of parallel control algorithms. Such nets can be efficiently analyzed; some properties can be decided for them in polynomial time, while for general-case nets such check requires in worst case exponential time (see for example [54,61,133,135,141,142]). Checking independency of transitions, producing stubborn and persistent sets is extremely simple for such nets - two transitions are dependent if and only if they belong to the same cluster; an enabled cluster constitutes a minimum stubborn set and a minimum persistent set.

Intersection of those two classes constitutes the class of α-nets [242] - a well-studied model used for specifying algorithm structure in PRALU.

It have been mentioned that the s-nets and especially the α-nets have some specific relations between their behavioral properties [110], and some of their properties can be efficiently decided [104, 109, 119, 124]. This section describes methods of their analysis and corresponding theoretical background. Some results from [105] are recalled; several mistakes have been corrected, new, more compact and structured proofs of most affirmations are given. New theoretical and experimental results are presented. Also, an interesting "by-product" has been obtained during research of these topics: it has been shown, that the basic stubborn set method can do more than it was supposed before - it detects at least one reachable marking, from which no deadlock is reachable, if such marking exists. Proof of this result is presented in Appendix A.

3.5.1 Properties and Analysis of α-Nets

Properties of α-Nets

In general case of Petri nets the three properties - liveness, safeness and reversibility - are independent from each other; in [172] examples of nets are given with all possible combinations of these properties. But for the subclasses considered here, certain dependence exists between these properties. In this subsection some results on that subject are presented.

The nets used for illustration are taken from [172] and [110]. If needed, the reachability graph is also shown.

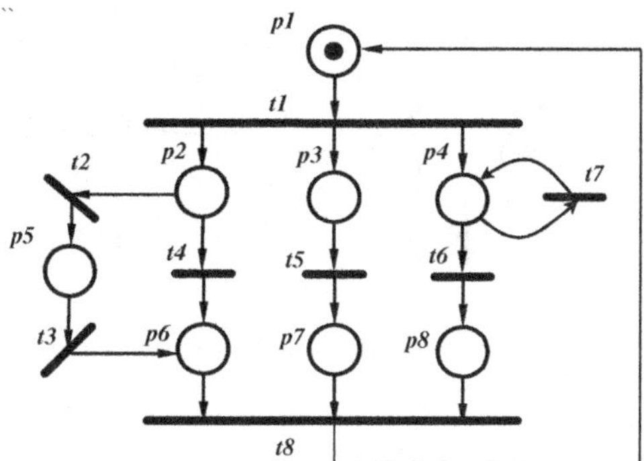

Fig. 3.10. A live and safe α-net

Below p_1 denotes the place marked in M_0 ($M_0(p_1) = 1$).

Lemma 3.26. *A live and safe s-net is reversible (Fig. 3.10).*

Proof. Suppose a live and safe s-net Σ is not reversible. Let M be a marking such that $M \in [M_0\rangle$ and $M_0 \notin [M\rangle$. A transition t such that $^\bullet t = p_1$ is live $\implies \exists M' \in [M\rangle$: $M'(p_1) > 0$. Σ is safe $\implies M'(p_1) = 1$. Σ is not reversible $\implies \exists p_i (i \neq 1) : M'(p_i) = 1$. $M' \geq M_0 \implies$ the firing sequence σ such that $M_0 \sigma M'$ is allowed in M', $M' \sigma M''$ and $M''(p_i) = 2$. This contradicts the assumption that Σ is safe.

Lemma 3.27. *[172] If a connected Petri net is live and safe, then it is strongly connected.*

So we can see that certain coordination exists between liveness, safeness and reversibility.

Lemma 3.28. *If for an α-net $\Sigma = (P, T, F, M_0)$ $\exists M \in [M_0\rangle : M > M_0$, then the net is not reversible (Fig. 3.12).*

Proof. Suppose Σ is reversible. Let $M \sigma M_0$. Let $T' = \{t \in T :^\bullet t \neq \{p_1\}\}$. Without lost of generality we may suppose, that all transitions in T' are disabled in M (otherwise fire all such transitions in σ, which can be fired before firing transition t_1 such that $^\bullet t_1 = \{p_1\}$, and consider the resulting marking as M). Reorder σ by permuting the independent transitions to obtain sequence $\sigma' \sigma''$ such that σ' is the longest sequence enabled in M_0. Let $M \sigma' M'$, $M_0 \sigma' M''$. Suppose there is transition t enabled in M''. Then there should be transition

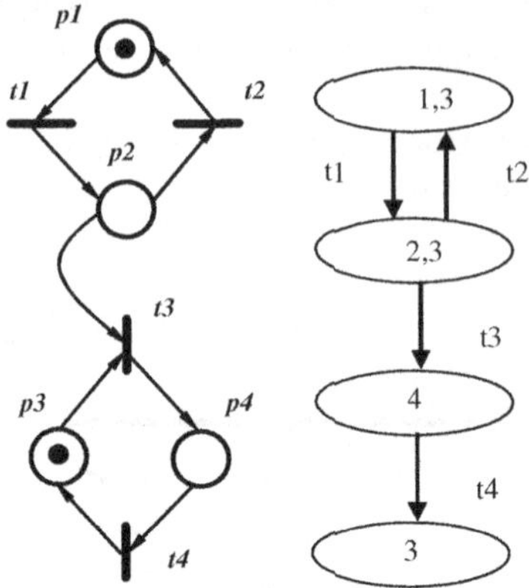

Fig. 3.11. A non-reversible Petri net and its reachability graph

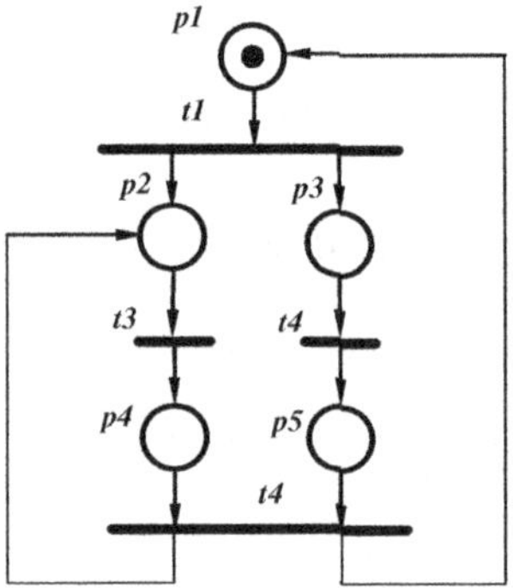

Fig. 3.12. A non-reversible and unsafe net (places p_2 and p_1 can be marked at the same marking)

$t' \in \sigma''$ removing tokens from $^\bullet t$ ($M' > M''$; even if $^\bullet t = \{p_1\}$, on the path from M' to M_0 should exist marking M''' such that $M'''(p_1) = 0$). As far as Σ is EFC, $^\bullet t =^\bullet t'$. Then t' is enabled in M'' and should belong to σ', not to σ''. It is a contradiction; hence $enabled(M'') = \emptyset$, $M'' \in [M_0\rangle$ is a deadlock, and Σ is not reversible.

Lemma 3.29. *If a live and strongly connected α-net $\Sigma = (P, T, F, M_0)$ is not safe, then $\exists M \in [M_0\rangle : M > M_0$.*

Proof. Let $M' \in [M_0\rangle$, $M'(p_i) > 1$. Let Q be a simple path in the net graph, leading from p_i to p_1 (Q exists, because Σ is strongly connected); let $V(Q)$ be the set of nodes on Q. Denote two of the tokens present in p_i at M' as a and b. Suppose, that if a transition $t \in V(Q)$ fires, and a is present in $p_j \in^\bullet t$, after its firing a is removed from p_j and added to its output place $p_k \in V(Q)$; otherwise, if b is present in $p_j \in^\bullet t$, after firing of t, b is removed from p_j and added to its output place $p_k \in V(Q)$. So, a and b by definition are situated in the places belonging to Q. Let $U \subset [M'\rangle$ be the set of markings for which a and b are defined. Let $M'' \in U$, a is in p_a, b is in p_b. Σ is live $\Longrightarrow$ there is a marking $M''' \in [M''\rangle$ such that $\{t_a, t_b\} \cap enabled(M''') \neq \emptyset$ (where $p_a \in^\bullet t_a$, $p_b \in^\bullet t_b$), $M''' \in U$. But Σ is EFC, hence $\{t'_a, t'_b\} \cap enabled(M''') \neq \emptyset$, where $p_a \in^\bullet t'_a$, $p_b \in^\bullet t'_b$ and $\{t'_a, t'_b\} \subset V(Q)$. Executing of t'_a or t'_b from M''' leads to a marking belonging to U, for which sum of distances between p_1 and the places, in which a and b are situated (calculated as number of corresponding arcs in Q), is less, than for M''' (and for M'). Applying such construction to this new marking we obtain, in finite number of steps, a reachable marking M such that a is in p_1 and b exists somewhere in the net, so $M > M_0$.

Lemma 3.30. *If a live and strongly connected α-net is not safe, then it is not reversible.*

Proof follows directly from lemmas 3.28 and 3.29

Lemma 3.31. *If an α-net Σ is strongly connected, reversible and has at least one live transition, then it is live (Fig. 3.10).*

Proof. Suppose Σ is not live. It is reversible $\Longrightarrow$ it has an initially dead transition t. Let p be a place such that $p \in^\bullet t$. The net is strongly connected, hence there is a path in the net graph from p_1 to p. Let t' be the last non-dead transition on that path. Without lost of generality we can suppose, that $p \in t'^\bullet$. (At least the first transition on the path is live, because at least one transition t'' such that $^\bullet t'' = p_1$ is live and any other transition with p_1 as its input place has no other input places (the net is EFC) and hence is live.) Then a marking M is reachable such that $M(p) > 0$. $M_0 \in [M\rangle$ and $M_0(p) = 0 \Longrightarrow$ to reach M_0 from M, a transition which has p as an input place, has to fire. But if that transition is live, then t is also live (because they share the same set of input places), which contradicts to the assumption that t is dead .

Lemma 3.32. *If an α-net Σ is strongly connected and reversible, then it is safe.*

Proof. If Σ has no live transitions, it is safe. If it has a live transition, from Lemma 3.31 Σ is live. Then, according to Lemma 3.30, it is safe.

Now we have all the necessary affirmations to prove the main theorem of this subsection.

Theorem 3.33. *An α-net is live and safe, if and only if it is reversible, strongly connected and has a live transition.*

Proof. Follows directly from Lemmas 3.26, 3.27 ($\Longrightarrow$); 3.31, 3.32 ($\Longleftarrow$).

Analysis of α-nets

There exist several efficient methods to decide liveness and safeness of EFC-nets and α-nets (for example, the reduction methods and the methods using linear algebraic approach [47, 60, 141, 248]). Some of them allow deciding some important properties such as liveness and safeness in polynomial time. But if the net under consideration turns to be incorrect, these methods are unable to localize breaches of correctness. At most, the reduction techniques allow finding the "bad" markings (such as deadlocks), but without obtaining firing sequences leading to them from initial marking. For practical applications such, as verification of software or hardware systems, this information would be very useful. Methods based on constructing reduced reachability graphs allow to perform such a profound analysis; and, as it will be shown below, for the α-nets the stubborn set method allows to decide whether the net is live and safe.

Theorem 3.33 describes dependency between reversibility, liveness and safeness of α-nets. So it is enough to check reversibility (and strong connectedness, which can be performed in linear time - see, for example, the algorithm described in [48]). On the other hand, if a non-live or unsafe α-net is not reversible, it would be useful to get for such net a firing sequence, leading to a marking from which it is impossible to return to the initial one. A method is proposed below, which allows doing that by constructing a subgraph of the reachability graph.

Theorem 3.34. *Let* $\Sigma = (P, T, F, M_0)$ *be a strongly connected* α-net, $G = (V, E)$ - *its full reachability graph*, $G_R = (V_R, E_R)$ - *its RRG created with the stubborn set method. G is strongly connected, if and only if G_R is strongly connected.*

Proof. If all transitions of Σ are dead in M_0, then the theorem evidently holds. Suppose there is an initially live transition in Σ.

$\implies$ Consider a net $\Sigma' = (P', T', F', M_0')$ such that: $P' = P \cup \{p_d\}$; $T' = T$; $\forall t \in T' : ({}^\bullet t' = {}^\bullet t;$ *if* $p_1 \notin t^\bullet$ *then* $t'^\bullet = t^\bullet$, *else* $t'^\bullet = t^\bullet \backslash \{p_1\} \cup \{p_d\})$, $M_0' = M_0$. Informally, p_1 is replaced by 2 places, one (p_d) is only an input place for some transitions, another is only an output place; the rest of the net remains as in Σ. Let $G' = (V', E')$ and $G_R' = (V_R', E_R')$ be full and reduced reachability graphs of Σ' correspondingly, where G_R' is constructed so, that for every M such that $M \in V_R'$ and $M \in V_R$, the same enabled cluster is selected for exploration, as when constructing G_R. If G is strongly connected, then Σ is reversible, and from Theorem 3.33 Σ is live and safe. Then the only reachable marking of Σ such that $M(p_1) > 0$ is M_0. Hence Σ' has the only deadlock M_d such that $M_d(p_d) = 1$, $M_d(P' \backslash \{p_d\}) = 0$, and $\forall M \in [M_0'\rangle(M_d \in [M\rangle)$; $V' = V \cup \{M_d\}$; $V_R' = V_R \cup \{M_d\}$. By construction, G_R is a subgraph of G. Let $M \in V_R$. Then $M \in V_R'$, and there is a path in G' from M to M_d. From Theorem 3.9 there is a path from M to M_d in G_R' and hence a path from M to M_0 in G_R. Hence G_R is strongly connected.

$\impliedby$ Suppose G is not strongly connected, G_R is strongly connected. Let M' be such marking of Σ that $M' \in [M_0\rangle$, $M_0 \notin [M'\rangle$. Then $M' \notin V_R$; in G on the path from M_0 to M' there is $M \in V_R$ (all enabled transitions in an initial marking of an α-net share an input place and constitute the stubborn set, so for any path in G starting from M_0, at least its first arc is present in G_R). Let $M\sigma M'$. If a transition t, belonging to the stubborn set selected for M, exists in σ, move it to the beginning of σ obtaining sequence $t\sigma_1$; if there is no such transition, add it to the beginning of σ. Let MtM_1. In the first case let $M_1' = M'$; t is independent of all transitions preceding it in σ, hence $t\sigma_1$ is enabled in M, and $M_1\sigma_1 M_1'$. Tn the second case let $\sigma_1 = \sigma$, let M_1' be such marking that $M'tM_1'$, $M_1\sigma_1 M_1'$ (from Lemmas 3.6 and 3.7 it exists). In both cases $M_1 \in V_R$, $M_0 \notin [M_1'\rangle$, $M_1' \notin V_R$. Applying this construction repeatedly, we will have at each step $M_i\sigma_i M_i'$, $M_i \in V_R$, $M_i' \notin V_R$, $|\sigma_i| \leq |\sigma_{i-1}|$. If at certain step σ_i will be reduced to 0 length, there will be a contradiction ($M_i = M_i'$ at the same time belongs and does not belong to V_R). If for every possible choice of t in the selected stubborn set, there is such step j that $\forall i > j(\sigma_i = \sigma_j)$, then the first transition in σ_j is ignored at M_j (and at all markings reachable from M_j in G_R). Then M_0 is not reachable from M_j in G_R, hence G_R is not strongly connected.

It follows from Theorem 3.34, that the stubborn set method allows to check reversibility of an α-net.

Theorem 3.35. *A strongly connected α-net is live and safe, if and only if its RRG is strongly connected and has more than one node.*

Proof. Apply Theorem 3.33 and Theorem 3.34.

Theorem 3.36. *Let $\Sigma = (P, T, F, M_0)$ be a strongly connected α-net, let $G_R = (V_R, E_R)$ be its RRG, let $M \in V_R$. If $M \in V_R$ and in G_R there is no path from M to M_0, then $M_0 \notin [M\rangle$.*

Proof. Suppose $M_0 \in [M\rangle$, $M\sigma M_0$ ($Mt_1 M_1 t_2 M_2 ... t_n M_0$). Let T_S be the cluster (stubborn set), selected for M. If $t_1 \in T_S$, then $M_1 \in V_R$. Suppose $t_1 \notin T_S$. As far as $M_0(P) = M_0(p_1) = 1$, $\forall p \in P \setminus \{p_1\}(M(p) > 0 \Rightarrow (\exists t_i \in \sigma \; p \in {}^\bullet t_i))$. Then $\exists t_j \in T_S \, (t_j \in \sigma, t_1 ... t_{j-1} \notin T_S)$. Then from Lemma 3.6 $Mt_j M_1' t_1 ... t_{j-1} M_j$; $M_1' \in V_R$. In both cases there is a marking $M' \in V_R$ such that $(M, M') \in E_R$, $M'\sigma' M_0$ and $|\sigma'| = |\sigma| - 1$. Applying this construction $|\sigma|$ times, we obtain a path in G_R from M to M_0. When such path does not exist, $M_0 \notin [M\rangle$.

Theorems 3.35 and 3.36 show, that the stubborn set method allows to decide whether an α-net is well-formed, to obtain a path from the initial marking to a non-desirable one (if such markings exist) and so, in certain respect, to verify correctness of a parallel system such as a parallel algorithm, structure of which corresponds to an α-net, and to localize a fault if it exists.

The complete algorithm of analysis of an α-net Σ is given below.

Algorithm 3.37

1. Check whether the net is strongly connected. If it is not, go to 6.
2. Construct a RRG G_R for Σ:
   a) Introduce the initial marking M_0 as a node and tag it "new".
   b) While "new" markings exist, do the following:
      i. Select a new marking M.
      ii. While there exist enabled transitions in M, select a cluster containing enabled transitions and do the following for each transition t in this cluster:

        A. Obtain the marking M' that results from firing t at M.

        B. If there is no node M', introduce M' as a node and tag M' "new".

        C. Introduce an arc with label t from M to M'.

        D. If M' is not safe, go to 8.

      iii. Remove label "new" from M.

3. If G_R is strongly connected, go to 7.
4. If there is in G_R a deadlock M_d, find in G_R a path from M_0 to M_d; find σ such that $M_0 \sigma M_d$; go to 10.
5. Find in G_R a marking M from which there is no path to M_0 in G_R; find σ such that $M_0 \sigma M$; go to 11.
6. Σ is not live or not safe; go to 12.
7. Σ is live and safe; go to 12.
8. Find in G_R a path from M_0 to the unsafe marking M_{us}; find σ such that $M_0 \sigma M_{us}$.
9. Σ is not safe; σ is a firing sequence leading from M_0 to an unsafe marking. Go to 12.
10. Σ is not live; σ is a firing sequence leading from M_0 to a deadlock. Go to 12.
11. Σ is not live or not safe; σ is a firing sequence leading from M_0 to a marking M such that $M_0 \notin [M\rangle$.
12. The end.

The expensive step of Algorithm 3.37 is its step 2 (the construction of the reduced state space), which requires in worst case exponential time and memory. An experimental evaluation of its efficiency will be presented below.

Examples and Discussion

Consider the examples of SFCs shown in Fig. 2.5; first of them is correct, others are not. Structure of all examples can be represented by the α-nets (Fig. 3.13).

The RRG constructed as a result of applying Algorithm 3.37 to the net from Fig. 3.13a is shown in Fig. 3.14b. For comparison, full reachability graph of the same net is shown in Fig. 3.14a. Grayed are the nodes of the full graph, present in the RRG. RRG is strongly connected, and the net turns to be live and safe.

Fig. 3.15 shows RRGs for the nets from Fig. 3.13b,c. First of those graphs is infinite, so only its part is shown, built until an unsafe marking is not detected (as in Algorithm 3.37). It is easy to see that the graphs are not strongly connected, and it is easy to find a firing sequence leading to an unsafe marking (Fig. 3.15a) or a deadlock (Fig. 3.15b; for example, $t_1 t_2 t_3 t_4 t_5$).

Main advantage of the proposed method is that it allows not only to decide liveness and safeness, but also to find a path to a non-desired state and in that sense to localize the fault, if it exists. Main disadvantage of the method is that in general case it does not decide liveness independently from safeness. Of course it is not the case, when an unsafe marking (Fig. 3.15a) or a deadlock (Fig. 3.15b) is detected; but if an ignoring occurs, we can only say that the net is not live or not safe. It can be illustrated by a simple example.

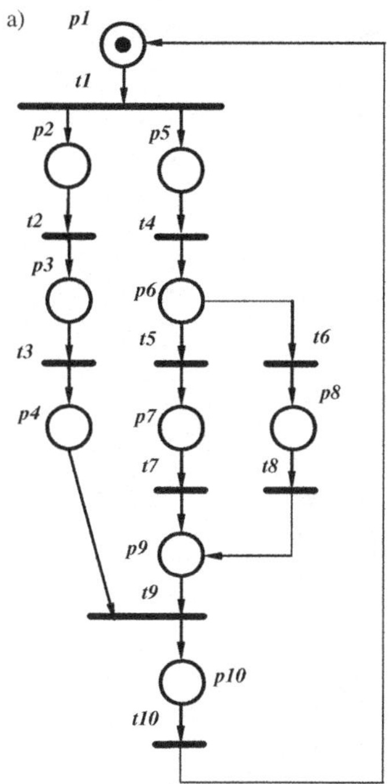

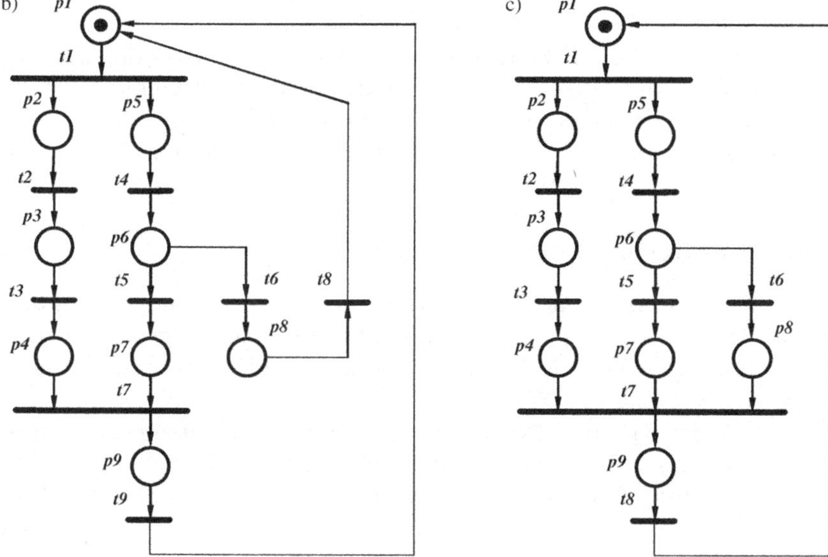

Fig. 3.13. α-nets corresponding to SFCs from Fig. 2.5

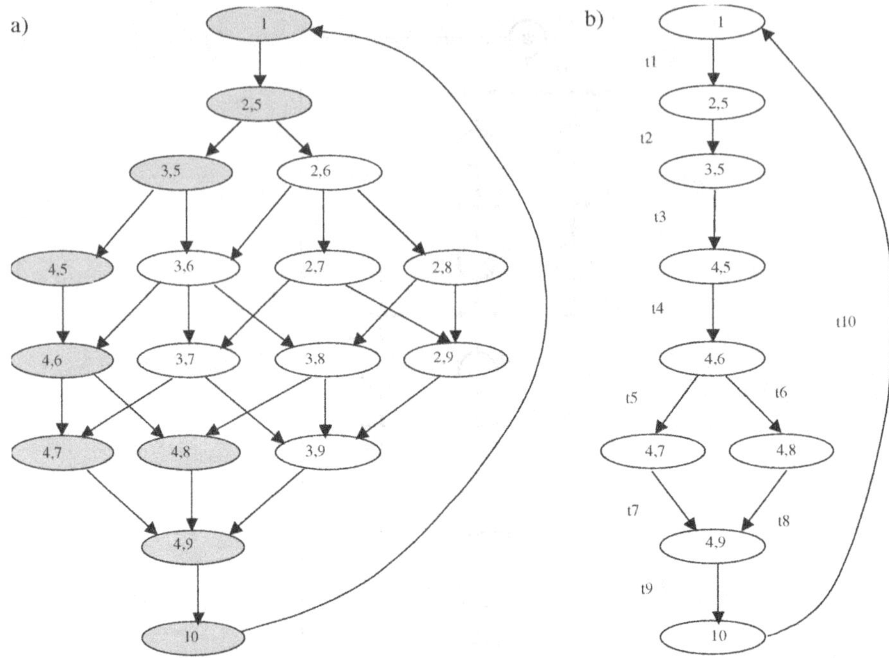

Fig. 3.14. Full reachability graph (a) and RRG (b) for the net from Fig. 3.13a

The net of Fig. 3.16a is safe, but not live. The net of Fig. 3.16b is live, but not safe. And the net of Fig. 3.16c is neither live nor safe. But applying Algorithm 3.37 (if numbering of the clusters corresponds to the numbering of transitions), for all the three the same graph will be obtained (Fig. 3.17).

In such cases the method always detects non-reversible markings, as follows from Theorems 3.35 and 3.36.

Experimental Results

For experimental evaluation of the method effectiveness, analysis of randomly generated α-nets has been performed, using full reachability graph construction and Algorithm 3.37. The algorithm for generation of pseudorandom nets was implemented according to [185].

In Table 3.1 the experimental results are shown. Net parameters are presented as $m \times n$, where m is number of places, n is number of transitions. The average percentage of time cost of RRG construction with respect of the time cost of the full graph construction is given for 10 randomly generated α-nets for every considered combination of parameters. The bottom row presents average percentage.

The arithmetic average of time cost percentage for all given examples is 63.9%, but as far as the gain increases when the full reachability graph becomes larger (the largest reachability graphs considered were constructed for the nets with parameters 72 × 40, and for such nets Algorithm 3.37 demonstrates maximal

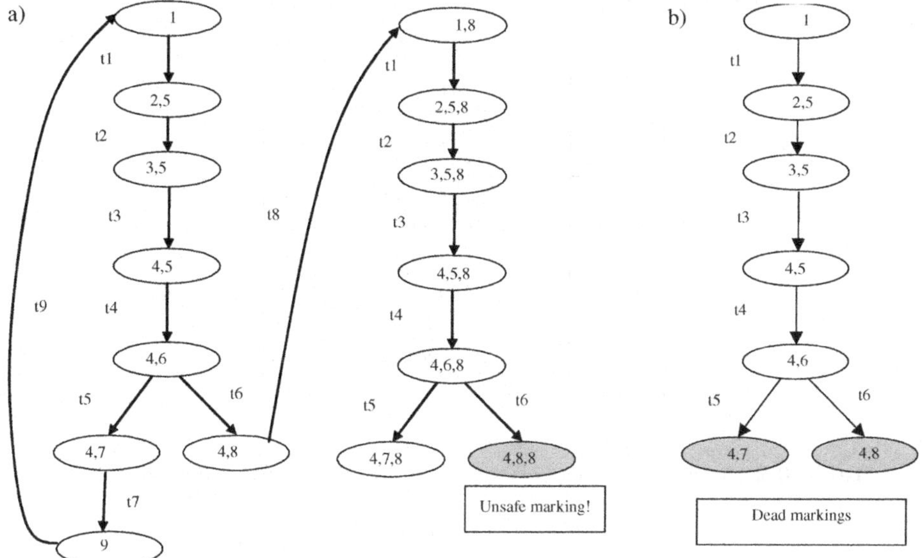

Fig. 3.15. RRGs for the nets from Fig. 3.13b (a) and Fig. 3.13c (b)

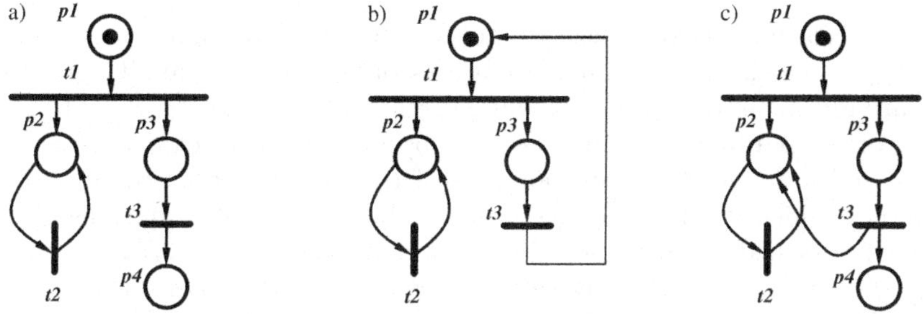

Fig. 3.16. Nets representing different breaches of correctness

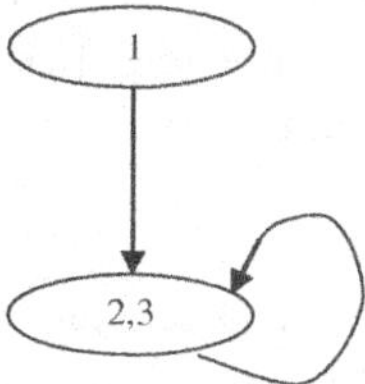

Fig. 3.17. RRG of the nets from Fig. 3.16

time saving), it makes sense to calculate average time costs also in the different way - from total time amount of analysis of all considered examples for each method. Such calculation brings result 42.5% for Algorithm 3.37.

Table 3.1. Results of experiments with Algorithm 3.37

net param.	Algor. 3.37 (%)	net param.	Algor. 3.37 (%)
12×20	55.8%	72×40	13.1%
36×20	38.4%	84×20	94.0%
36×40	89.8%	84×40	17.3%
36×60	92.1%	96×20	90.3%
48×40	74.4%	96×40	54.3%
60×20	94.6%	96×60	78.9%
60×40	65.8%	108×60	28.0%
72×20	71.7%	120×60	63.6%
average:	**72.8%**		**54.9%**

Table 3.2. Average numbers of arcs of full reachability graphs and RRGs

	net parameters				
	48×10	60×20	72×30	84×40	96×50
full graph	10.5	21.6	34.5	67.7	86.6
RRG	9.7	17.4	27.2	31.9	38.1

For the relation between size of full and reduced reachability graphs, see Table 3.2, in which average numbers of arcs of the full reachability graphs and RRGs are given. Reducing of size and, correspondingly, construction time with growing of net parameters, which can be observed for some examples, is caused by increased number of interconnections of the nets for some combinations of parameters, which in turn may decrease number of reachable markings.

Other performed experiments also show that Algorithm 3.37 allows to save about 50% of average analysis time and space for the nets with up to 140 reachable markings. With growing size of the full reachability graph, gain of Algorithm 3.37 in comparison with full reachability graph construction grows. But, of course, Algorithm 3.37 remains exponential in the size of the net. If time or memory cost of the algorithm turns to be too high, other methods should be used, such as the net reduction method, briefly described in Section 2.3 (allowing to decide liveness and safeness, and - by building reachability graph for the reduced net - also to obtain a "bad" marking, but not a firing sequence leading to a "bad" marking. For details see [201, 235, 249]). Experiments demonstrate, that for all examples considered the net reduction method works quicker (49.4% of time of full reachability graph construction, calculated as average value for all examples; 29.2%, if calculated from total time amount of analysis of all considered examples). Of course, the net reduction method also requires less memory.

3.5.2   A Hypothesis on EFC-Nets

Hypothesis: A strongly connected EFC-net is live and bounded, if and only if its RRG is finite and every transition of the net occurs in every terminal component of the RRG.

If this is true (and it turns to be true for many examples we checked), then the stubborn set method directly applied to an EFC-net allows to decide whether it is live *and* bounded. Structure of the proof of similar result obtained for α-nets, presented in subsection 3.5.1, does not seem to be applicable for this more general case, because it is essentially based on the fact that the initial marking of an α-net is single-token.

The hypothesis mentioned above specifies an interesting direction of further research.

3.5.3   Analysis of s-Nets

Lemma 3.26 demonstrates importance of s-nets - single-token initial marking is something more than just a particular restriction of initial state. A live and safe s-net is a model of a cyclic system, which starts its operation from single initial state and returns eventually to the same state. This is how almost any engineering object (and, at certain abstraction level, a natural object) can be described.

Below an affirmation (Theorem 3.41) is presented, showing that liveness of a bounded s-net can be decided by means of the stubborn set method. Theorem 3.41 was proven in [114], but the proof presented below is new and much shorter. We start with some preliminaries.

Lemma 3.38. *[74, 212] Ignoring does not occur in an RRG, if the net is bounded and strongly connected.*

Lemma 3.39. *[217] If in a RRG no ignoring occurs, then a transition t is live, if and only if it is live in the RRG.*

Lemma 3.40. *[217] If in a RRG no ignoring occurs, then if the full reachability graph is finite and contains a terminal component TC, then the RRG contains a terminal component TC_R such that $TC_R \subseteq TC$, and a transition t occurs in TC_R if and only if it occurs in TC; in the reverse direction, each terminal component TC_R of RRG is a subgraph of some terminal component TC of the full reachability graph such that t occurs in TC_R if and only if it occurs in TC.*

Theorem 3.41. *A bounded and strongly connected s-net $\Sigma = (P, T, F, M_0)$ is live, if and only if its RRG $G_R = (V_R, E_R)$ is strongly connected and every transition $t \in T$ occurs in it.*

Proof. $\implies$ According to Lemma 3.38, there is no ignoring in G_R. Consider a terminal component TC_R of G_R (it exists, because Σ never reaches a deadlock).

TC_R is strongly connected by definition. According to Lemma 3.40, there is terminal component TC in the full reachability graph with the same set T' of occurring transitions. Σ is live $\implies T' = T$. Then $\exists t \in T' :^\bullet t = \{p_1\}$, and there is $M \geq M_0$ in TC_R. Σ is bounded, hence $M = M_0$. Then $G_R = TC_R$, G_R is strongly connected and every transition $t \in T$ occurs in it.

$\impliedby$ If G_R is strongly connected and every transition occurs in it, then there is no ignoring in G_R. Hence Lemma 3.39 can be applied, and according to it every transition $t \in T$ is live, so Σ is live.

Theorem 3.41 shows, that by means of the stubborn set method in its basic form it is possible to decide liveness of the bounded Petri nets with single-token initial marking. It is generally considered that the basic stubborn set method needs some modifications to be able to verify liveness properties [212, 215].

Example

Below an example of applying the analysis approach, based on Theorem 3.41, to a parallel real-time algorithm for a controller is shown. The algorithm is taken from [9]. It is intended to control a chemical reactor. The reactor is fed after start signal with two kinds of liquids from measuring vessels, which feed

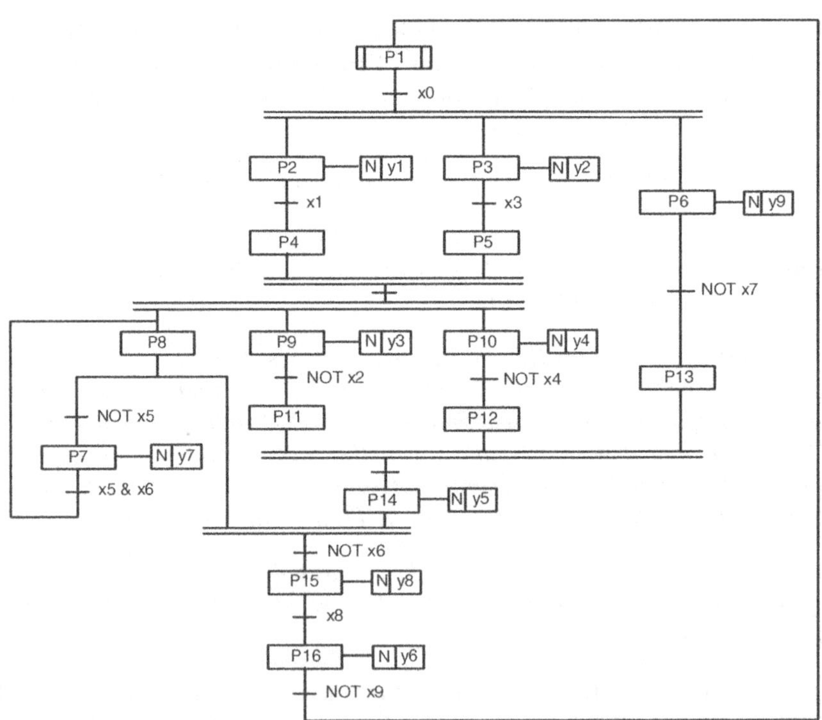

Fig. 3.18. A controller program in SFC

from the storage vessels. When reaction between the liquids is completed, the reactor is discharged into catch vessel. When the reactor is empty, the process product is transported to the storage vessel using a carriage. To ensure complete reaction the liquid in the reactor is agitated by a stirrer. For details see [9].

In Fig. 3.18 the SFC control algorithm is shown, Fig. 3.19 presents the corresponding Petri net. Note that the net does not belong to the class of EFC nets and it is an s-net. Full reachability graph of this net contains 29 nodes and 58 arcs (Fig. 3.20).

A reduced reachability graph (one of possible versions) is shown in Fig. 3.21. It has only 12 nodes and 13 arcs. It is easy to see that the graph is strongly connected and contains an arc for every transition of the net. It follows from Theorem 3.41 that the net is live (supposing that it is bounded). It means that the algorithm will never attain such a state that part of its code becomes unreachable (and that initially there is no unreachable code). In this respect the algorithm is correct.

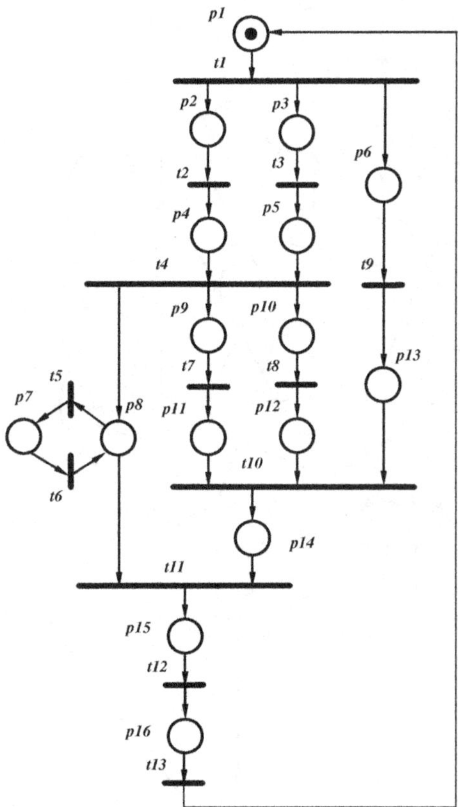

Fig. 3.19. Petri net corresponding to SFC shown in Fig. 3.18

3.6 Minimization of Space

Methods of lazy state space constructions, such as described above, evidently reduce not only time, but also space of analysis algorithms (in comparison to the full state space exploration). Theoretically there is no necessity of huge memory

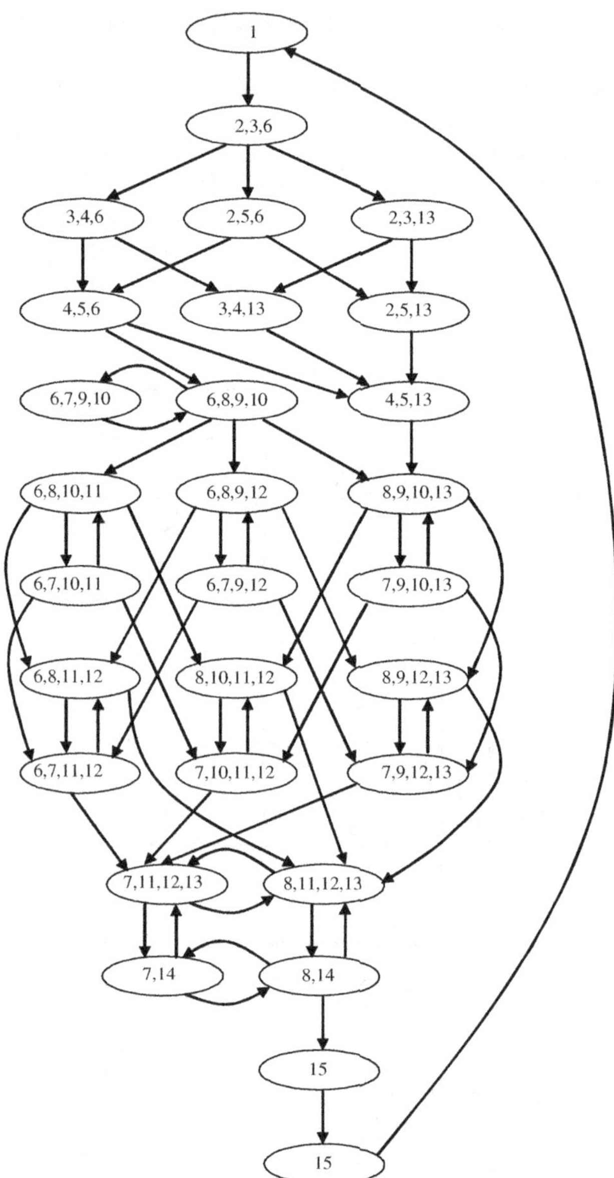

Fig. 3.20. Full reachability graph of the net shown in Fig. 3.19

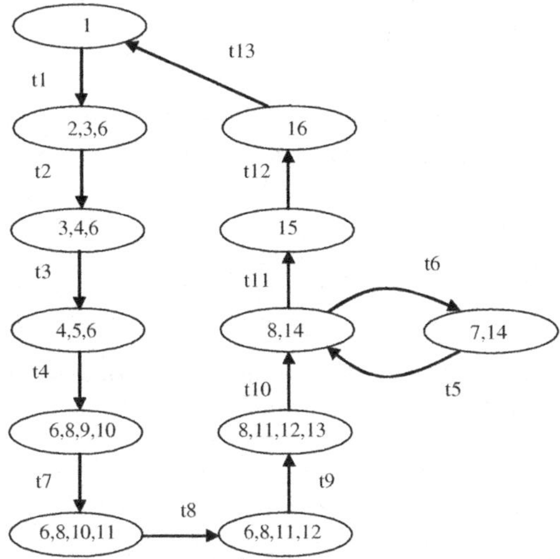

Fig. 3.21. RRG of the net shown in Fig. 3.19

to solve Petri net analysis problem; there is a polynomial-space algorithm of deadlock detection in a safe Petri net, but it is absolutely inapplicable because its time consumption is woeful. Generally, all known algorithms solving verification tasks in relatively small memory are extremely slow [217].

In this section we discuss a less radical approach, reducing memory, keeping it however exponential in the worst case. The approach is based on removing from memory some of the intermediate states. Some results are recalled from [112, 118, 123, 131]; several mistakes are corrected.

3.6.1 Dynamic Reduction of Reachability Graphs

Consider the problem of deadlock detection. There is no necessity to keep in memory the whole (even reduced) reachability graph with all intermediate (non-deadlock) markings. It is also evident that throwing away all these states may cause eternal looping (if the reachability graph has cycles). Therefore some of intermediate markings should be kept. Which ones? The following lemmas [112] give the answer (or at least one of the possible answers).

Lemma 3.42. *If the graph of a Petri net Σ is acyclic, then the corresponding reachability graph Σ is acyclic.*

Proof. If the graph of Σ is acyclic, then it specifies a partial order on the set of places. Assign to every place p a number $n(p)$ according to this order, so that for every place p and transition $t \in^\bullet p$: $n(p) > \Sigma_{p' \in \bullet t} n(p')$. Let $f(M) = \Sigma(M(p_i)n(p_i))$. Then for any t $MtM' \Rightarrow f(M) < f(M')$. Then no two markings can be mutually reachable, and reachability graph of the net has no cycles.

Lemma 3.43. *For every cycle L in the reachability graph of a Petri net there is a cycle L_Σ in the net graph such that every transition belonging to L_Σ marks an arc in L.*

Proof. Let σ be a firing sequence corresponding to a cycle in the reachability graph $(M\sigma M)$. Suppose that the transitions belonging to σ constitute no cycles in the net graph. Then $\exists t \in \sigma\; \exists p \in^\bullet t\; \forall t' \in \sigma\; p \notin t'^\bullet$. Then $M\sigma M' \Rightarrow M(p) > M'(p) \Rightarrow M \neq M'$ - a contradiction.

Lemma 3.44. *Let $M\sigma M'$, M and M' are comparable[7]. Then there is a cycle L_Σ in the net graph such that every transition belonging to L_Σ belongs to σ.*

Proof is analogous to the proof of Lemma 3.43 (should be repeated twice: for the cases $M > M'$ and $M < M'$).

Lemma 3.45. *Let $M\sigma M$. Then every transition $t \in \sigma$ belongs to a cycle of the net graph, such that all transitions of this cycle belong to σ.*

Proof. Suppose the opposite. Then $\exists t \in \sigma\; ((\exists p \in^\bullet t\; \forall t' \in \sigma\; p \notin t'^\bullet) \vee (\exists p \in t^\bullet\; \forall t' \in \sigma\; p \notin^\bullet t'))$. Then $M\sigma M' \Rightarrow M(p) \neq M'(p) \Rightarrow M \neq M'$ - a contradiction.

The algorithm presented below is a modification of the well-known algorithm of reachablilty graph construction for a Petri net $\Sigma = (P, T, F, M_0)$.

Algorithm 3.46

1. Introduce M_0 as a node and tag it "new". $B := \{M_0\}$.
2. Select $K \subseteq T$ such that for every cycle in the net graph at least one transition belongs to K.
3. While "new" markings exist, do the following:
   a) Select a new marking M.
   b) If no transitions are enabled in M, tag M "deadlock".
   c) While there exist enabled transitions in M, do the following for each enabled transition t in M[8] :
      i. Obtain the marking M' that results from firing t from M.
      ii. If there is no node M', introduce M' as a node and tag M' "new".
      iii. Introduce an arc (M, M'), labelled by t, and tag M' "new".
      iv. If on a path from M_0 to M there exists a marking M'' such that $M > M''$, then communicate "The net is unbounded" and go to 4.
      v. If $t \in K$, add M' to B.
   d) If M is not a "deadlock" and $M \notin B$, do the following:
      i. For every pair of arcs (a, M) and (M, b) draw an arc (a, b).
      ii. Remove node M with all its incident arcs.
   e) Else remove label "new" from M.
4. The end.

---

[7] The corresponding relation is defined in subsection 2.1.2.

[8] As Algorithm 3.46a, the variant of Algorithm 3.46 will be referred, in which step 3c has the form: "While there exist enabled transitions belonging to T_S at M, do the following for each enabled transition $t \in T_S$".

Theorem 3.47. *Algorithms 3.46 and 3.46a stop for every Petri net.*

Proof. Let $G = (V, E)$ be the graph constructed by the algorithm. Suppose that the net is bounded and the algorithm never stops. That means that there exists a cycle in the reachability graph such that every marking of it is added to V and then removed, and the loop never stops. From Lemma 3.43 it follows, that at least one of these markings will be included in set B and never deleted from V. Hence it cannot be tagged as "new" more than once, and such eternal looping is impossible - a contradiction.

Suppose that the net is unbounded and the algorithm never stops. Then the algorithm considers new markings (never considered before) infinitely often (another possibility of looping is excluded by Lemma 3.43). Any "long enough" firing sequence leading to new markings will go through markings M and M' such that $M' > M$, which follows from the fact that P is finite. Then, as follows from Lemma 3.44, a marking M'' between them will be added to B and never removed from V. That means that graph G will grow infinitely.

It follows that sooner or later there will be a "long enough" path in G such that there will exist some markings M and M' in it, such that $M' > M$. Then the algorithm will detect unboundedness (step 3c.iv) and stop.

The proof is completed; note that it is valid for both variants of the algorithm.

Theorem 3.48. *Algorithms 3.46 and 3.46a detect all deadlocks of a Petri net or its unboundedness.*

Proof follows from the algorithm description and Theorem 3.9.

Note that execution of step 2 of Algorithm 3.46 is a non-trivial task, if we want to minimize cardinality of set K (which would reduce the needed memory amount). It can be formulated as a task of *minimal decyclization* of an oriented graph or of finding *minimum feedback arc set* [158]. This task has applications in electrical engineering, design of discrete devices, scheduling and so on. The topic is discussed in Appendix B.

3.6.2   Reducing the Space for Various Analysis Tasks

Application of modified Algorithm 3.46 to solving some other analysis tasks is discussed below.

- **Boundedness.** It is easy to see, that n-boundedness can be decided by Algorithm 3.46, if number of tokens in the places is checked on-the-fly. Algorithm 3.46 also detects unboundedness, as follows from Theorem 3.48.
- **Liveness.** Algorithm 3.46 in its basic form cannot decide liveness. Consider the following modification: add to step 3d.i "label arc (a, b) by all transitions labelling (a, M) and (M, b)". Let this variant be denoted as Algorithm 3.46b. Graph built by Algorithm 3.46b allows liveness checking of any bounded net - it is enough to check, whether every transition marks an arc in every terminal component.

- **Reversibility.** A bounded net is reversible, if and only if graph G constructed by Algorithm 3.46 for this net is strongly connected.
- **Reachability, coverability and conservatism.** Algorithm 3.46 by construction considers every reachable marking of a bounded Petri net, so the properties mentioned above can be checked by means of it (with slight evident modification).
- **Firing sequences.** To check whether, for given net and given initial marking, firing sequence σ is possible, Algorithm 3.46 should be modified by forbidding removing from the graph the markings belonging to a path, which corresponds to an initial subsequence of σ (unless it is certain that this subsequence cannot be continued forming σ).

3.6.3   Example

Consider the Petri net shown in Fig. 3.13c. Its reachability graph is shown in Fig. 3.22. It has 13 nodes. In Fig 3.23 graph G is shown; Fig. 3.23a presents the graph at a stage of Algorithm 3.46 execution, when its number of nodes is maximal (assuming search in BFS order); Fig. 3.23b presents its final form. Maximal number of nodes of G is 6; every marking has been considered only once. The situation differs if the state space is searched in DFS order; then maximal number of nodes is also 13, but some markings have been considered 2 or even 3 times.

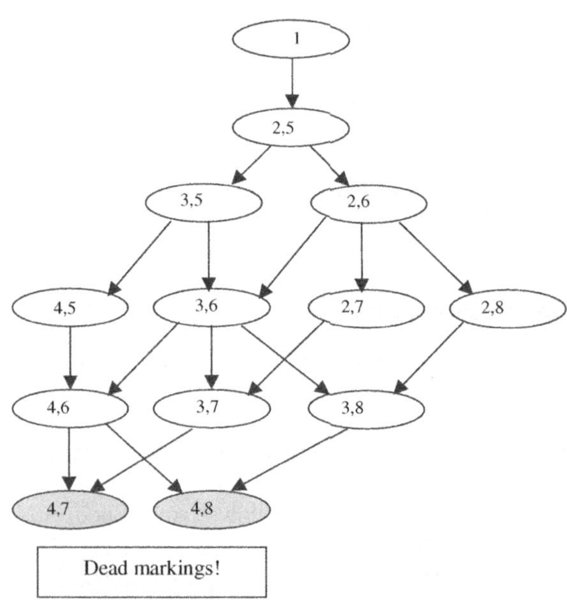

Fig. 3.22. Reachability graph of the net shown in Fig. 3.13c

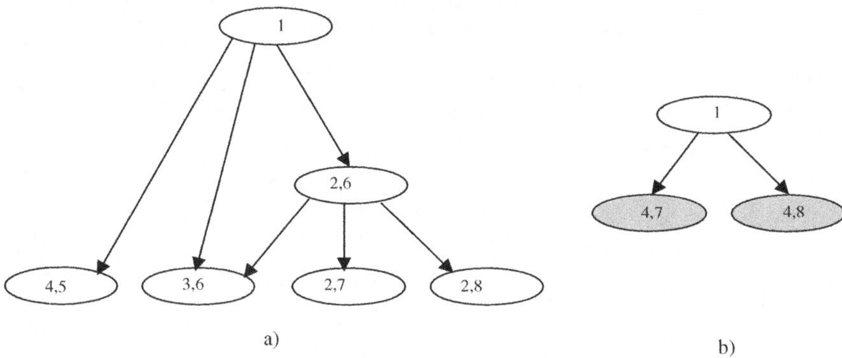

Fig. 3.23. Intermediate (a) and final (b) graph G constructed by Algorithm 3.46 for the net shown in Fig. 3.13c

It is also interesting to compare RRG built using the stubborn set method and graph G' built according to Algorithm 3.46a for this example. RRG contains 7 nodes, G' has maximally 3 nodes. Similar situation arises with combination of the proposed approach and maximal concurrent simulation (such modification of Algorithm 3.46 is not described here): graph (let it be G'') has maximum 3 nodes, when the graph created by maximal concurrent simulation has 5 nodes.

3.6.4 Conclusive Notes on the Method

How much do we gain (and loose) when using the method proposed above?

Exact analytical evaluation of its space- and time- complexity turns to be a difficult task even for the nets with very restricted structure, and we do not have sufficient evidence to decide how good it is generally. There are lot of "good" and "bad" examples. Some general notes on the complexity of the method are presented below.

- Gaining in memory, we loose in time. In most cases the described algorithms re-calculate some of the states several (sometimes many) times.
- BFS search order seems to be preferable here. DFS allows in many cases to save more space, but it leads to multiple re-calculation of markings and increases time consumption drastically.
- Combination of the method with the stubborn set method is efficient, because the stubborn set method, by avoiding interleaving, reduces radically the number of paths leading to a node in the reachability graph. So, re-calculation of the states occurs rarely for such combination.

The space-complexity is reduced here at the expense of time-complexity, and practical using of this approach makes sense in the cases when memory size is more critical parameter than time. The approach seems to be especially efficient for deadlock detection and reachability analysis and, generally, analysis of safeness properties. It is compatible not only with the stubborn set method, but also with maximal concurrent simulation (which implies in fact BFS search order, and, as we have seen, it is preferable here).

The methods described in Chapter 4 (except Section 4.2) also avoid keeping in memory all investigated markings, however using a different approach.

4. Decomposition for Analysis

A perspective approach to analysis of Petri nets and, generally, of parallel discrete systems, is based on the idea of decomposition. The task of net analysis is reduced to the task of analysis of the blocks of its decomposition, which may be considerably smaller than the net itself. That procedure can simplify analysis of large nets. Decomposition turns to be useful also for the synthesis purposes. More about theory and applications of net decomposition can be found in [15, 19, 36, 102, 169, 249, 251].

4.1 Block Decomposition

In this section the method of block decomposition and analysis, developed by A. Zakrevskij [244, 249, 251], is briefly described. This information is necessary for understanding next sections. New results are presented in subsections 4.1.3 (modification of the method for analysis of cyclic nets) and 4.1.4 (experimental results). The main idea of the method is following: a net is decomposed into blocks, and every block is analyzed separately; a partial order relation between the blocks is specified, according to which the blocks can be analyzed. The method is intended first of all for analysis of a special class of nets called the *operational Petri nets* [244, 251].

4.1.1 Operational Petri Nets

Definition 4.1. *A Petri net $\Sigma = (P, T, F)$ is an operational Petri net (OPN), if it has nonempty sets of input ($P^{in} = {}^{\bullet}T \setminus T^{\bullet}$) and output ($P^{out} = T^{\bullet} \setminus {}^{\bullet}T$) places.*

It is easy to see, that Definition 4.1 is similar to Definition 3.12, with the difference that a block is defined as a subnet, unlike an operational Petri net. In both cases for any transition t belonging to an OPN or a block, $P^{in} \cap t^{\bullet} = \emptyset$, $P^{out} \cap {}^{\bullet}t = \emptyset$.

Definition 4.2. *An initial marking M_0 is correct for an operational Petri net $\Sigma = (P, T, F)$, if the following conditions are satisfied.*

A. Karatkevich: Dynamic Analysis of Petri Net-Based Discrete Sys., LNCIS 356, pp. 63–85, 2007.
springerlink.com

1. $M_0(P \setminus P^{in}) = 0$.
2. Σ is safe for M_0.
3. From any marking $M \in [M_0\rangle$, a deadlock is reachable.
4. All reachable deadlocks are the terminal markings (in the deadlocks only the output places contain tokens).

In Fig. 4.1 an example of operational PN is shown. The correct initial markings for this net are $\{p_1, p_2\}$ and $\{p_3\}$.

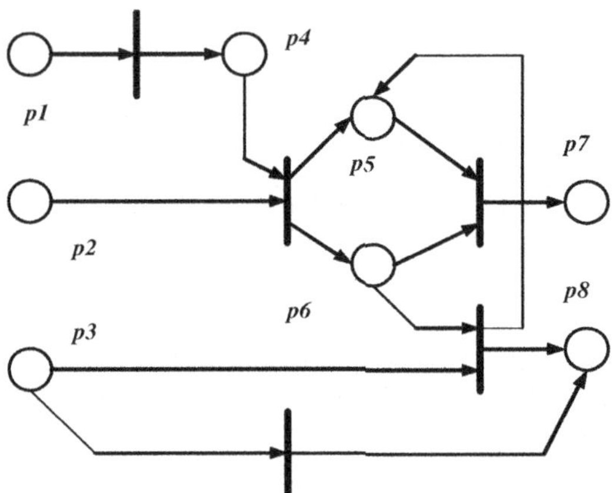

Fig. 4.1. An example of operational Petri net

Union of the nets Σ_1 and Σ_2 such that $P_1^{in} \cap P_2^{in} = \emptyset$ and $P_1^{out} \cap P_2^{out} = \emptyset$ will be called their *composition*. For such nets only those places can be common, which are input for one net and output for the other. In Fig. 4.2 (taken from [249]) possible variants of composition of two operational PNs are shown.

For the theory of operational Petri nets (properties of correct initial markings, relations that can be implemented by such nets, the net decomposition) see [249, 251]

An operational PN can be used as a specification of a structure of parallel algorithmalgorithm!parallel, which is initiated by some correct input state and terminates attaining one of the output states. In particular, such net may have only one input place and one output place. Such two-pole structures are typical for computational and control algorithmsalgorithm!logical control [249].

4.1.2   Analysis of Operational Petri Nets

Block decomposition of a Petri net can be performed using Lemma 3.14 [249,251].

For the analysis of an operational PN its decomposition into blocks should be first performed. It can be made by finding transitive closure of the relation of alternative joint on the set of transitions (see Definition 3.13 and Lemma 3.14). Then in the oriented graph corresponding to relation R on the set of blocks, where $(\Sigma_1, \Sigma_2) \in R \Leftrightarrow P_1^{out} \cap P_2^{in} \neq \emptyset$, all the cycles should be united (it can be

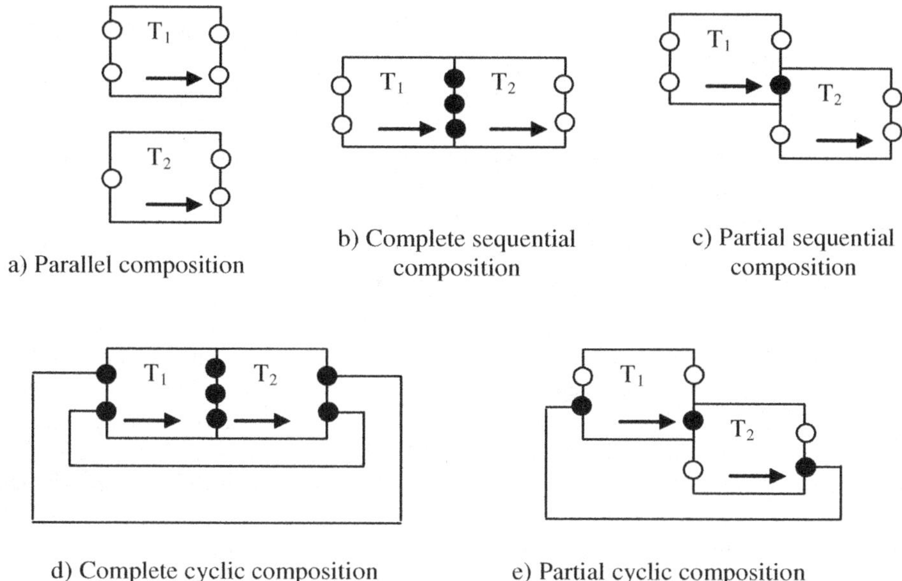

a) Parallel composition

b) Complete sequential composition

c) Partial sequential composition

d) Complete cyclic composition

e) Partial cyclic composition

Fig. 4.2. Variants of composition of two operational Petri nets

performed in linear time, using the algorithm of detection of the strongly connected components [48]), i.e. all blocks belonging to the same strongly connected component should be composed into one block (step 2 of Algorithm 4.3). After that R becomes a relation of partial order. Then it specifies the order of the net analysis: a block can be analyzed only when all the blocks, with which it shares its input places, have already been analyzed. The main task of analysis of operational PN is to check whether the given initial marking is correct and, if it is correct, to obtain all terminal markings reachable from it. Block decomposition allows analyzing operational nets using the following algorithm given in [251]. Let $\Sigma = (P, T, F, M_0)$ be an operational PN.

Algorithm 4.3

1. If $M_0(P \setminus P^{in}) > 0$, go to 8.
2. Decompose Σ into minimal blocks.
3. Calculate the relation of partial order R.
4. Unite all the cycles composed by blocks.
5. $B := \{M_0\}$.
6. While $\exists M \in B\ M(P \setminus P^{out}) > 0$, do:
   a) Select a block Σ_i such that $\exists M \in B\ M(P_i^{in}) > 0$ and there is no block Σ_j not analyzed yet for which $(\Sigma_j, \Sigma_i) \in R$.

    b) For each marking $M \in B$ such that $M(P_i^{in}) > 0$, find the terminal markings of Σ_i reachable from it and replace M in B by those markings (tokens outside Σ_i do not change their positions).[1]

    c) If at least one of the initial markings for Σ_i is found to be incorrect, go to 7.

7. The initial marking M_0 is correct, and B contains all the terminal markings reachable from it in Σ. The end.

8. The initial marking is incorrect. The end.

4.1.3 Analysis of a Class of Cyclic Nets

A cyclic Petri net satisfying some conditions can be "unfolded" into an operational PN. This transformation can be easily performed if the initial marking contains a single token (see Fig. 3.10 and Fig. 3.1). This condition can be generalized - if all transitions, such that their input places are marked in the initial marking, share the same set of input places, then the net can be transformed into an operational PN. The same occurs, if the similar condition holds for the sets of *output* places. It is easy to see, that if and only if the original net is live and safe, the resulting operational PN will be quasi-live and will have only one correct initial marking and one terminal marking.

So, if a Petri net is supposed to be reversible, and if it satisfies the condition formulated above, then it can be transformed into an operational PN and analyzed by the method described. Such analysis would allow deciding liveness and safeness of the original net. Also, if the net is not live or not safe, it would allow to obtain an unsafe marking or marking from which the net cannot return to the initial one, together with firing sequence leading to such marking, or the list of dead transitions.

It follows that next algorithm (taken from [251], with some modifications) can be used for liveness and safeness analysis of a net, such that all tokens can be removed from the places containing them in the initial marking (or put there) only by a single transition firing. That means in particular, that it is applicable to analysis of α-nets. The algorithm can be easily extended in such way that for a non-LS net it will allow to localize its faults, as described above. But for such localization all explored paths in the reachability graph should be remembered, otherwise it is enough to keep in memory the set D.

Let $\Sigma = (P, T, F, M_0)$ be a PN satisfying the above condition.

Algorithm 4.4

1. Add to P a new place for each place marked in M_0. Replace all edges in the net graph, leading to a place marked in M_0, by the edges leading to the corresponding new place (from the same transitions). An operational PN Σ' is obtained. The places marked in M_0 are its input places, and the new places are its output places.

---

[1] In [249, 251] it is supposed, that the blocks are analyzed by constructing their full reachability graphs. Possible combination of block decomposition and the stubborn set method is considered in Section 4.3.

2. Check the initial marking M_0 for Σ' using Algorithm 4.3.
3. If M_0 is found to be incorrect:
   a) if from a reachable marking no terminal marking is reachable:
      i. if such marking is reachable that all output places are marked together with some internal places, then Σ is not safe;
      ii. else Σ is not live;
   b) if an unsafe place is found, then Σ is not safe.
4. If M_0 is found to be correct:
   a) if a terminal marking is found such that not all output places are marked or if some transitions have not fired during simulation, then Σ is not live;
   b) else Σ is live and safe.
5. The end.

4.1.4 Example and Experimental Results

Consider the net shown in Fig. 3.10; it can be transformed into an operational PN (step 1 of Algorithm 4.4) (Fig. 3.1). In Fig. 4.3 block decomposition of the net is shown.

In Fig. 4.4 the RRG is shown, constructed de facto by Algorithm 4.3 for the net shown in Fig. 3.1. The full reachability graph of this net has 14 nodes and 32 edges (see Fig. 3.2).

In Fig. 4.5, 4.6 the analysis of the net similar to the net shown in Fig. 3.1, but with the reachable dead non-terminal markings, is shown.

For the experiments a random net generator [185] and the program implementing the described method were used. For the largest nets investigated - with 65 transitions and 120 places - maximum time of analysis was 3 ms (the experiments were performed on a PC with processor K6 II). As far as it is

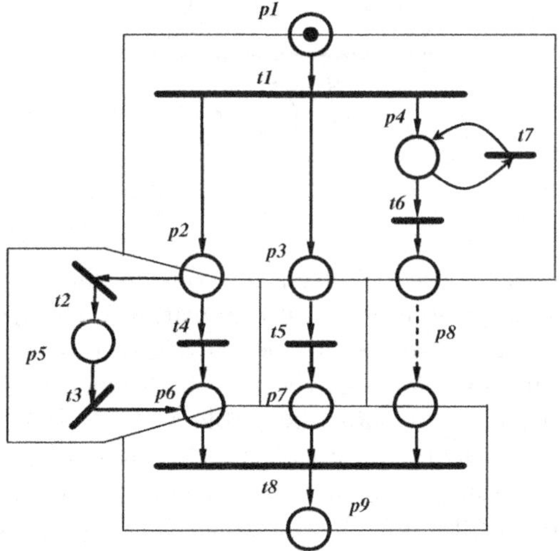

Fig. 4.3. Block decomposition of the net shown in Fig. 3.1

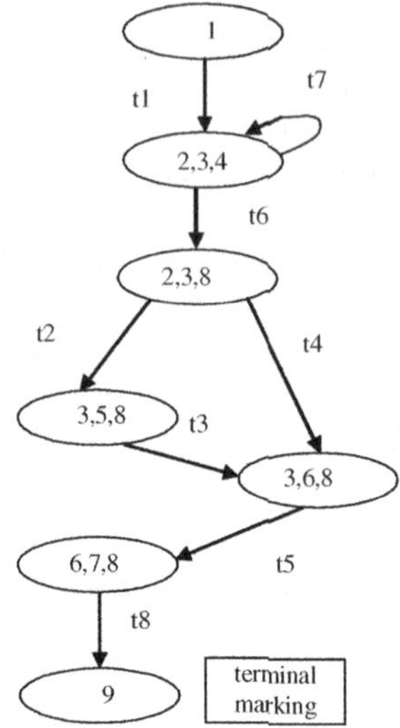

Fig. 4.4. RRG for the net shown in Fig 3.1

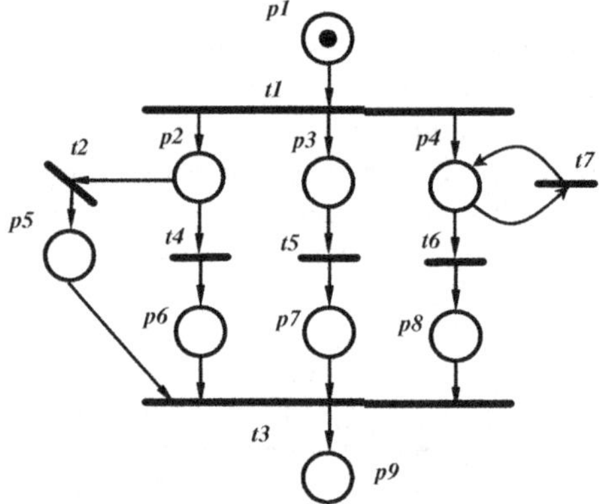

Fig. 4.5. A net with non-terminal dead markings

supposed that the blocks are analyzed by constructing the full reachability graphs, most critical parameter of net decomposition is the size of largest block. The size of a block means here the number of transitions in it. Average results (size of the largest minimal blocks of the nets with given parameters) obtained in the series of experiments are shown in Table 4.1.

Results of another series of experiments (with different set of nets) are shown in Table 4.2. Here we compare number of markings explored by the presented method with number of all reachable markings. The experiments demonstrate, that the method explores about 40% of the state spaces of the considered nets.

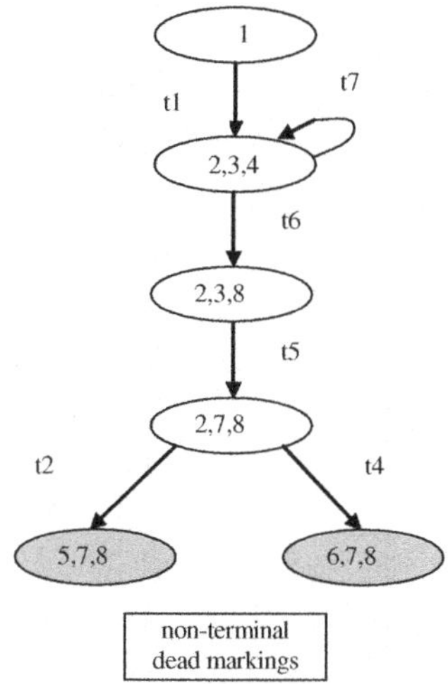

Fig. 4.6. RRG for the net shown in Fig. 4.5

Table 4.1. Average number of transitions in largest blocks after the net decomposition

	transitions		
places	20	40	60
20	16	35	
40	18	7	21
60	18	10	6
80	18	17	7
100	19	21	5
120			6
average:	17.8	18.0	9.0

Next conclusion follows from these results: with growth of the net, the relation between its size and the size of its largest block increases. This means that the method is efficient for large nets. It is less efficient for the nets with $|P| << |T|$; it is intuitively clear that the structure of such nets prevents deep decomposition.

4.2 Hierarchical Decomposition

A Petri net model in its original form has no hierarchical structure. But, on the one hand, hierarchical structures are required in many applications of Petri nets in system engineering. On the other hand, decomposition is useful for reducing complexity of many synthesis and analysis tasks. In many publications hierarchical net models and the ways of hierarchical decomposition of nets of different kinds have been proposed (see, for example, [13, 37, 62, 79, 100, 101, 136, 146, 175, 179, 191, 214]).

Table 4.2. Reduction of state space by the block decomposition method

net №	reachable markings	explored markings	percentage
1	49	11	22.45%
2	66	23	34.85%
3	43	13	30.23%
4	19	7	36.84%
5	47	10	21.28%
6	11	11	100.00%
7	26	13	50.00%
8	10	7	70.00%
9	126	31	24.60%
10	51	16	31.37%
11	29	9	31.03%
12	141	41	29.08%
average:	**51.5**	**16**	**40.14%**

A hierarchical structure is easy to deal with if all elements at all levels are of the same kind and need no special ways of description. Below we describe an algorithm that allows to transform an ordinary, live and safe Petri net into a hierarchical one. The method proposed, being a combination of hierarchical and block decomposition, is rather simple; on the other hand, it allows selecting the subnets of complex structures. Such decomposition simplifies reachability, liveness and deadlock analysis of the nets.

4.2.1 A Conception of Hierarchical Decomposition of Petri Nets

If we are going to introduce a hierarchical Petri net, then of course each level of the model will consist of Petri nets and all we have to do is to define dependencies between the nets of different levels, adding maybe some necessary restrictions. In the approach we use, the places of a higher-level net may correspond to the lower-level nets.

For example (see Fig. 4.7): the high level of a hierarchical net corresponding to the flat (not hierarchical) net shown in Fig. 3.19 is presented [8]. *Macroplace* mp_1 corresponds to a subnet, specified by transitions t_{12}, t_{13}; mp_2 corresponds to t_2; mp_3 - to t_3, and so on[2].

In such system some processes may take place "inside" a macroplace. Another possible variant is a net with *macrotransitions* (see for example [79, 100, 191]); we believe, that it is less convenient for our purposes, because it contradicts the conception of immediate transition firing.

Let us consider properties of a subnet that can be replaced by a place. Let $\Sigma' = (P', T', F')$ be a subnet of $\Sigma = (P, T, F)$. First of all, Σ' must have an

---

[2] Another possibility of decomposition of this net is selecting a macroplace corresponding to transitions t_2, t_3 and a macroplace corresponding to t_7 and t_8. See [8].

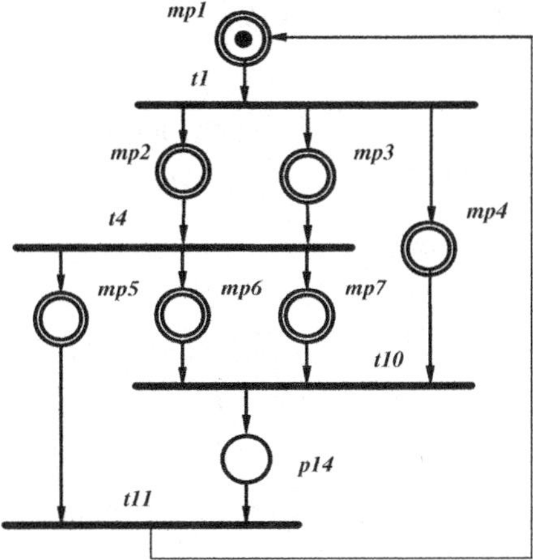

Fig. 4.7. A net with macroplaces corresponding to the net from Fig. 3.19

interface - a set of places $P'^{i/o}$ incident to some external transitions. $P'^{i/o}$ has two subsets (maybe intersecting) - P'^{in} (input places) and P'^{out} (output places), consisting of the places that are output and input for the external transitions, correspondingly. The following condition must be satisfied:

$$\forall t \in (T \setminus T')$$
$$((t^\bullet \cap P'^{in} \neq \emptyset) \Rightarrow (t^\bullet \supseteq P'^{in})) \wedge ((^\bullet t \cap P'^{out} \neq \emptyset \Rightarrow (^\bullet t \supseteq P'^{out})). \quad (4.1)$$

It means that a subnet can get the tokens from outside only in all of its input places simultaneously, and it can loose tokens from all its output places by an external transition firing also only simultaneously. To guarantee this, the initial marking should also be taken into account. It implies one more necessary condition:

$$(\exists p \in P' \ M_0(p) > 0) \Rightarrow (p \in P'^{in} \wedge (\forall p' \in P'^{in} \ M_0(p') = M_0(p))) \vee$$
$$(p \in P'^{out} \wedge (\forall p' \in P'^{out} \ M_0(p') = M_0(p))). \quad (4.2)$$

Denote by M'_{in} the marking such that $M'_{in}(p) = 1$ if and only if $p \in P'^{in}$ and by M'_{out} the marking such that $M'_{out}(p) = 1$ if and only if $p \in P'^{out}$. It is reasonable to consider M'_{in} as an initial marking for the subnet ($M'_{in} = M'_0$). The following condition is also important:

$$\forall M : (M \in [M'_{in}\rangle) \Rightarrow (M'_{out} \in [M\rangle). \quad (4.3)$$

And finally, if all places in P'^{out} obtain tokens, no tokens should be allowed to remain "inside" the subnet. This condition is described by the following formula:

$$\forall M : (M \in [M'_{in}\rangle) \Rightarrow (M \not> M'_{out}). \quad (4.4)$$

Definition 4.5. *[107] A P-block is a subnet Σ' of Petri net Σ, satisfying conditions (4.1-4.4)[3].*

If a subnet Σ' of Petri net Σ is a P-block, let us construct the net $\Sigma_H = (P_H, T_H, F_H)$ (where H denotes the higher level; $mp \in P_H$ is the macroplace):

$$
\begin{aligned}
P_H &= (P \setminus P') \cup \{mp\}; \\
T_H &= (T \setminus T'); \\
F_H &= \{(x,y)|(((x \in P_H \wedge y \in T_H) \vee (x \in T_H \wedge y \in P_H)) \wedge \\
&\quad (x,y) \in F) \vee ((x = mp) \wedge (y \in T_H) \wedge (\exists p \in P' \, (p,y) \in F)) \vee \\
&\quad ((y = mp) \wedge (x \in T_H) \wedge (\exists p \in P' \, (x,p) \in F))\}.
\end{aligned}
$$
(4.5)

Correspondence between markings of Σ_H and Σ, when Σ is safe, is described by next formula:

$$
M_H(p) = \begin{cases} M(p), & p \in P \\ max(M(p')), p' \in P', \, p = mp. \end{cases}
$$
(4.6)

Lemma 4.6. *Let Σ be a safe Petri net, Σ' its P-block, Σ_H constructed according to (4.5), M' is a projection on Σ' of marking M of Σ. Then for Σ $M_1 \in [M_2\rangle$, if and only if:*

1. *for Σ_H: $M_{1H} \in [M_{2H}\rangle$, $M_{1H}(mp) = M_{2H}(mp) = 0$, or*
2. *for Σ_H: $M_{1H} \in [M_{2H}\rangle$, $M_{1H}(mp) = 0$, $M_{2H}(mp) = 1$; for Σ': $M'_{out} \in [M'_2\rangle$; or*
3. *for Σ_H: $M_{1H} \in [M_{2H}\rangle$, $M_{1H}(mp) = 1$, $M_{2H}(mp) = 0$; for Σ': $M'_1 \in [M'_{in}\rangle$; or*
4. *for Σ_H: $M_{1H} \in [M_{2H}\rangle$, $M_{1H}(mp) = M_{2H}(mp) = 1$; for Σ': $M'_1 \in [M'_2\rangle$, if the firing sequence in Σ_H leading from M_{2H} to M_{1H} does not contain transition t such that $mp \in {}^\bullet t$, or ($M'_{out} \in [M'_2\rangle$, $M'_1 \in [M'_{in}\rangle$), otherwise; or*
5. *$M_{1H} = M_{2H}$, $M'_1 \in [M'_2\rangle$.*

Proof. $\Longleftarrow$ Let σ_H be a firing sequence in Σ_H such that $M_{2H}\sigma_H M_{1H}$, σ' a firing sequence in Σ' such that $M'_{in}\sigma'M_{out}$.

1. If $\exists t_1 \in \sigma_H : mp \in {}^\bullet t_1$, then before t_1 in σ_H there is transition t_2 such that $mp \in t_2^\bullet$ (because $M_{1H}(mp) = M_{2H}(mp) = 0$). Insert σ' into σ_H before t_1; replace t_1 by transition t''_1 such that ${}^\bullet t''_1 = ({}^\bullet t_1 \setminus \{mp\}) \cup P'_{out}$, $t''^\bullet_1 = t_1^\bullet$;

---

[3] The essential differences between P-blocks and Zakrevskij's blocks (see Definition 3.12) are the following: (1) sets of input and output places of a P-block may intersect; (2) those sets are defined for P-blocks only by their external relations; in Zakrevskij's blocks their internal relations are also taken into account; (3) Definition 3.12 does not contain condition (4.1) or an equivalent condition; (4) a P-block can have only one correct initial marking. There are subnets satisfying both definitions (the two-pole blocks, for example), but the notions of blocks and P-blocks do not cover each other.

replace t_2 by transition t_2'' such that $t_2''^{\bullet} = (t_2^{\bullet} \setminus \{mp\}) \cup P_{in}'$, $^{\bullet}t_2'' = {}^{\bullet}t_2$. If $\exists t_3 \in \sigma_H : mp \in (^{\bullet}t_3 \cap t_3^{\bullet})$, then insert σ' into σ_H before t_3; replace t_3 by transition t_3'' such that $^{\bullet}t_3'' = (^{\bullet}t_3 \setminus \{mp\}) \cup P_{out}'$, $t_3''^{\bullet} = (t_3^{\bullet} \setminus \{mp\}) \cup P_{in}'$. If there are more transitions in σ_H satisfying the condition $(mp \in {}^{\bullet}t) \vee (mp \in t^{\bullet})$, then make the corresponding changes as described above. The sequence thus obtained leads from M_2 to M_1 in Σ. If $\forall t \in \sigma_H : mp \notin {}^{\bullet}t$, then $M_2 \sigma_H M_1$ in Σ.

2. In this case there is $t_1 \in \sigma_H$ such that $mp \in {}^{\bullet}t_1$. Replace t_1 with t_1'' as described above. Insert at the beginning of σ_H the sequence leading from M_2' to M_{out}'. If there are more transitions in σ_H satisfying the condition $(mp \in {}^{\bullet}t) \vee (mp \in t^{\bullet})$, then make the changes as in item 1. The obtained sequence leads from M_2 to M_1 in Σ.

3. In this case there is $t_2 \in \sigma_H$ such that $mp \in t_2^{\bullet}$. Replace t_2 with t_2'' as described in item 1. Insert at the end of σ_H the sequence leading from M_{in}' to M_1'. If there are more transitions in σ_H satisfying the condition $(mp \in {}^{\bullet}t) \vee (mp \in t^{\bullet})$, then make the changes as in item 1. The sequence thus obtained leads from M_2 to M_1 in Σ.

4. If, on the path from M_{2H} to M_{1H} in Σ_H, there exists a transition t such that $(mp \in {}^{\bullet}t) \vee (mp \in t^{\bullet})$, then act as in items 2 and 3, obtaining a firing sequence leading from M_2 to M_1 in Σ. Else $M_2 \sigma_H \sigma'' M_1$ in Σ, where $M_2' \sigma'' M_1'$ (firing sequence $\sigma_H \sigma''$ is enabled in M_2, because in this case all transitions in σ_H belong to T and are independent of all transitions in σ'').

5. In this case evidently $M_2 \sigma' M_1$ in Σ.

$\implies$ Let σ be the sequence such that $M_2 \sigma M_1$ in Σ. Replacing in the sets of input and output places of transitions subsets P_{in}' and P_{out}' by $\{mp\}$ and removing all transitions that do not belong to T_H, we obtain the sequence σ_H such that $M_{2H} \sigma_H M_{1H}$ in Σ_H (if $M_{1H} = M_{2H}$, then $|\sigma_H| = 0$). The removed transitions constitute one or more enabled sequences for Σ', that lead from M_2' to M_{out}', from M_{in}' to M_1', from M_{in}' to M_{out}' or from M_2' to M_1' (all possible situations are described in items 1-5).

The following three theorems are the main results of this subsection; they demonstrate how the hierarchical decomposition simplifies the Petri net analysis. Theorems 4.7 and 4.8 relate to liveness and deadlock analysis, correspondingly; Theorem 4.9 describes dependency between cardinality of reachability sets of decomposed net and the elements of decomposition.

Theorem 4.7. [4] *Let Σ be a safe Petri net, Σ' its P-block, Σ_H constructed according to (4.5). Σ is live, if and only if Σ_H is live and Σ' is quasi-live.*

Proof follows from Lemma 4.6.

---

[4] Theorem 4.7 is similar to the *Reduction theorem* from [191]; but in [191] the subnets corresponding to *modules*, being in fact the macrotransitions, belong to such restricted class of Petri nets as *marked graphs*. Another difference between our approach and the approach presented in [191] is that the decomposition proposed there is in principle two-level; our decomposition can be easily generalized for multiple levels.

Theorem 4.8. *Let Σ be a safe Petri net, Σ' its P-block, Σ_H constructed according to (4.5). Σ has a reachable deadlock M_d, if and only if Σ_H has a reachable deadlock M_{dH}. Then*

$$M_d(p) = \begin{cases} M_{dH}(p), & p \in P_H \\ 0, & p \in P' \setminus P'^{out} \\ M_{dH}(mp), & p \in P'^{out}. \end{cases} \tag{4.7}$$

Proof follows from Lemma 4.6.

Theorem 4.9. *Let Σ be a safe Petri net, Σ' its P-block, Σ_H constructed according to (4.5). Then*

$$|[M_0\rangle| = \tag{4.8}$$
$$= |\{M_H \in [M_{0H}\rangle | M_H(mp) = 0\}| + |\{M_H \in [M_{0H}\rangle | M_H(mp) = 1\}||[M'_{in}\rangle|,$$

where $[M_0\rangle$ concerns Σ, $[M_{0H}\rangle$ - Σ_H, $[M'_{in}\rangle$ - Σ', correspondingly.

Proof follows from Lemma 4.6.

For example, the reachability graphs of the net given in Fig 4.7 and the subnets corresponding to its macroplaces together have 19 nodes and 14 arcs. Compare it with 29 nodes and 58 arcs of the net from Fig. 3.19 (see Fig. 3.20). Such reduction is possible, because in the first case interleaving is completely avoided.

Suppose that we want to know, whether the marking $\{p_7, p_{14}\}$ is reachable in the net from Fig. 3.19. Searching directly the state space of this net, we obtain this marking after at least 9 steps (transition firings). Analyzing the hierarchical net (Fig. 4.7), we have to check whether the marking $\{mp_5, p_{14}\}$ is reachable (3 steps), and then whether in the subnet corresponding to mp_5 marking $\{p_7\}$ is reachable (1 step). It takes together only 4 steps.

Of course, a net can be decomposed into (or composed from) many subnets in hierarchical way; the affirmations analogous to presented above but describing multi-level decomposition can be easily proved by using induction. Let us call a net decomposition *P-decomposition*, if all the subnets, the net is decomposed into, are the P-blocks (except of the net at the highest level).

4.2.2   Properties of P-Decomposition

Lemma 4.10. *If the nets Σ' and Σ'' are P-blocks of net Σ and Σ'' is a subnet of Σ', then Σ'' is a P-block of Σ'.*

Proof follows from the definition of subnet and Definition 4.5.

This ensures the possibility of multi-level decomposition.

Lemma 4.11. *If the nets Σ' and Σ'' are P-blocks of net Σ and neither Σ'' is a subnet of Σ', nor Σ' is a subnet of Σ'', but $T' \cap T'' \neq \emptyset$, then $\Sigma_1 = (\Sigma' \setminus \Sigma'')$, $\Sigma_2 = (\Sigma' \cap \Sigma'')$, $\Sigma_3 = (\Sigma'' \setminus \Sigma')$ and $\Sigma_4 = (\Sigma' \cup \Sigma'')$ are the P-blocks of Σ.*

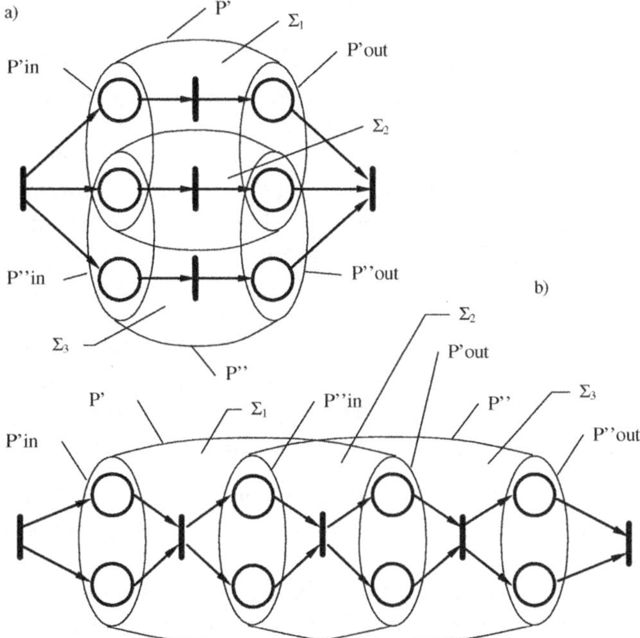

Fig. 4.8. Variants of intersection of P-blocks

The lemma can be proved by demonstrating, that there are only two possible kinds of intersecting of P-blocks (parallel and sequential, as shown in Fig. 4.8), and that for each kind of intersection result of corresponding operations on the P-blocks is a subnet satisfying (4.1-4.4). The complete proof is given in Appendix C.

It does not mean that for a Petri net there exists only one P-decomposition with maximal number of blocks, because the P-blocks may intersect by places, having no common transitions. It is easy to see, that all blocks Σ_1, Σ_2 and Σ_3 in Fig. 4.8b cannot be replaced all together by the macroplaces.

4.2.3 Finding P-Blocks

Of course, the most comfortable situation for analysis is when we deal with a system synthesized with the top-down approach, where the well-formed blocks correspond to the macroplaces [146, 191, 253]; then we have a "ready-made" decomposition. Otherwise we have to decompose the given net. As it is shown by Theorem 4.9, even finding some P-blocks can simplify the net analysis.

In order to check whether a subnet is a P-block, it is by definition enough to check conditions (4.1-4.4). Checking conditions (4.3,4.4) in general case requires constructing the reachability graph, which is a task of exponential time- and space-complexity; but if it is known that the net is live and safe, the checking can be performed easier.

Lemma 4.12. *If the net Σ is live and safe and a subnet Σ' of this net satisfies (4.1,4.2), then Σ' is a P-block.*

Proof. If Σ is live, then after firing of a transition t_1 such that $P'^{in} \subseteq t_1{}^\bullet$ a marking is reachable in which transition t_2 such that $P'^{out} \subseteq{}^\bullet t_2$ is enabled. It means that in Σ' $\forall M \in [M'_{in}) : (M' \in [M) : (\forall p \in P'^{out} : M'(p) > 0))$. As far as Σ is safe, $M'(p) = 1$. Let $M'_{in}\sigma M'$. Now it is enough to prove, that $M'(P' \setminus P'^{out}) = 0$. Suppose the opposite. Then consider marking M'_{add} in Σ' such that $M'_{add}(p) = 1$ if and only if $M'(p) > M'_{out}(p)$. If $\exists M'' \in [M'_{add}) : M''(P'^{out}) > 0$, then Σ is not safe. Otherwise, as far as Σ is live, there is marking $M_1 \in [M_0)$ such that $M_1 > M'_{add}$ and $M_1 > M'_{out}$, and there is marking $M_2 \in [M_1)$ such that $M_2 > M'_{add}$ and $M_2 > M'_{in}$. Then σ is enabled in M_2, and executing of σ from M_2 leads to an unsafe marking (the places marked in M'_{add} will be unsafe). So, supposing that $M'(P' \setminus P'^{out}) \neq 0$ we have come to a contradiction. Then conditions (4.3,4.4) hold for Σ', and Σ' is a P-block.

Even in a case when it is easy to decide whether a subnet is a P-block, decomposition is not a simple task, because the number of subnets exponentially depends on the net size.

One of the ways to found some of P-blocks is selecting in the Petri net the subnets being the SM-components (different variants of such approach can be found in [8, 17, 175]). The net shown in Fig. 4.7 corresponds to the macronet shown in Fig. 4 in [8] - all its macroplaces correspond to the SM-components. Of course, the P-blocks may have more complex structure; for detecting such blocks, another method is needed.

It is easy to see, that any P-block Σ' such that $P'^{in} = {}^\bullet T' \setminus T'^\bullet$, $P'^{out} = T'^\bullet \setminus {}^\bullet T'$ is a *block* in the sense of Definition 3.12 (let us call such blocks the $\mathfrak{P}$-blocks), but not every block is a P-block. Lemma 3.14 [249] provides a simple way to find for a given Petri net its partition into minimal blocks. This partition can serve as a base for finding a P-decomposition. Some blocks have to be united to obtain P-blocks.

We propose the following algorithm for finding P-blocks for a given live and safe Petri net Σ (a modified algorithm from [107]).

Algorithm 4.13

1. Find the partition of Σ into minimal blocks (according to Lemma 3.14).
2. For each block Σ' such that

$$\forall t \in T : ((t^\bullet \cap P'^{in} \neq \emptyset) \Rightarrow (t^\bullet \supseteq P'^{in})), \tag{4.9}$$

check conditions (4.1,4.2). If Σ' satisfies these conditions, it is considered as a P-block. If Σ' does not satisfy (4.2), continue with another block not checked yet. Else attach the blocks Σ'' such that $P''^{in} \subseteq P'^{out}$ expanding Σ' until it is possible and Σ' does not satisfy (4.1,4.2). If the expanded block satisfies these conditions, it is considered as a P-block.

3. For the blocks that are not included in P-blocks at the previous step consider combinations of their parallel composition satisfying (4.9) and process them as in step 2.

Theorem 4.14. *Algorithm 4.13 detects every minimal $\mathfrak{P}$-block of an LS-Petri net.*

Proof. Let Σ' be a minimal $\mathfrak{P}$-block. As far as minimal blocks specify a partition on a Petri net (Affirmation 5.14 from [249]), and a union of the blocks is a block (Affirmation 5.15 from [249]), Σ' is a minimal block or a union of minimal blocks. In the first case it will be detected immediately in step 2 of Algorithm 4.13. In the second case there is a block Σ_1 such that P'^{in} is the set of its input nodes, Σ_1 satisfies (4.9) and (4.2), but does not satisfy (4.1). Σ_1 is either a minimal block (then it will be processed by step 2), or a parallel composition of minimal blocks (then it will be detected in step 3 and then processed by step 2). Adding another minimal block Σ_2 to Σ_1, as described in step 2 of Algorithm 4.13, is possible in every case while $\Sigma_1 < \Sigma'$, and then $\Sigma_2 \subset \Sigma'$. So, in finite number of steps the situation will be attained such that $\Sigma_1 = \Sigma'$, and Σ' will be detected as a P-block, because it satisfies (4.1,4.2) and, as far as the net is live and safe, from Lemma 4.12 it also satisfies (4.3,4.4).

What will happen, if we apply Algorithm 4.13 to a net which is safe, but not necessarily live? Then the blocks it will detect, will of course satisfy Definition 3.12, but additional analysis will be necessary to check whether they satisfy (4.3,4.4). Such analysis can be performed for all those blocks in any order, because for each of them not more than one correct initial state is possible (and specified by its structure). Then, if all detected blocks turn to be the P-blocks, reachability (according to Lemma 4.6) and liveness (according to Theorem 4.7) can be decided for the net. It is easy to show, that if for any of the blocks condition (4.3) or (4.4) is not satisfied, the net is not live.

After applying Algorithm 4.13 to a flat net and composing the P-blocks into the macroplaces, the same algorithm can be applied to the highest-level net Σ_H, and so on, while the highest-level net is decomposable. In such a way a multi-level hierarchical net can be constructed in the bottom-up order. Some notes should be taken into account here: first, if some of the P-blocks detected by Algorithm 4.13 intersect by places, but have no common transitions, then, unlike for the *blocks*, partial composition is impossible for them; they may be only in a *complete composition* (see Fig. 4.2). In such cases the sum of cardinalities of reachability set of the P-blocks equals cardinality of reachability set of their composition, and considering those P-blocks as separate macroplaces seems to have no sense; for analysis purposes those P-blocks should be united. In particular, there is no sense to build a macroplace from an SM-component, if it can be built from a "larger" SM-component. So, to step 1 of Algorithm 4.13 the following statement can be added: "Unite all the complete sequential compositions of the blocks"; then there will be no guaranty of minimality of the P-blocks obtained, but it will simplify the net structure and the net analysis. On the other hand, there is no sense in our analysis to consider several P-blocks being in parallel composition as a single macroplace (as it is done in [8]).

4.3 Decomposition and Persistent Sets

One of the reasons why decomposition is useful for analysis is that it allows to "reuse" results of analysis of the particular blocks of a system. Analysis by constructing the state spaces usually does not allow such reuse and can lead to repetitive analysis of the same subsystems. See Fig. 4.9: here during simulation using stubborn sets, the block specified by transitions t_4, t_5 is simulated twice. In case of constructing the full reachability graph, this block and the block specified by t_2, t_3 would be simulated 3 times. Both decomposition methods described above avoid such duplication.

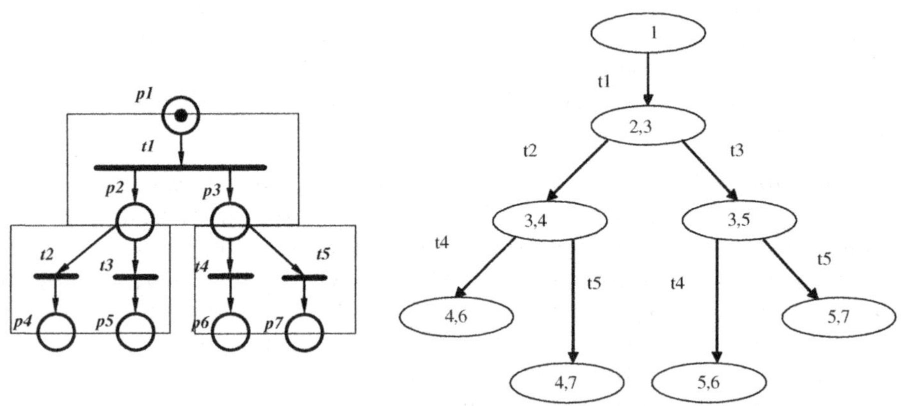

Fig. 4.9. A Petri net (with block decomposition) and RRG

On the other hand, for complete analysis of a block we have to know all its possible initial states. That is why certain limitations of applicability exist, and for this reason both decomposition methods are intended for the safe nets (which means limited number of initial markings of the blocks; the methods can be generalized for the bounded nets).

In spite of certain differences, the methods of analysis by means of decomposition can be considered as belonging to the family of persistent set methods [130], which follows from the next theorem.

Theorem 4.15. *Let* $\Sigma = (P, T, F)$ *be an operational Petri net,* $\Sigma_1 = (P_1, T_1, F_1)$ *a block in it, at marking M there are enabled transitions in Σ_1 and there is no such block Σ_2 in Σ, that in Σ_2 there are enabled transitions at M and $(\Sigma_2, \Sigma_1) \in \widehat{R}$ (where $\widehat{R}$ is a transitive closure of relation R defined in subsection 4.1.2). Then $T_1 \cap enabled(M)$ is a persistent set.*

Proof. T_1 contains an enabled transition; no enabled transition in T_1 has an input place in common with any transition outside T_1 (which follows from Definition 3.12); so, T_1 satisfies two of 3 conditions of a stubborn set (Definition 3.4). If it satisfies the remaining condition (every disabled transition in T_1 has an empty input place p such that all transitions in $^\bullet p$ are in T_1), then T_1 is a stubborn set and the statement holds. Suppose T_1 is not stubborn; then there is a disabled transition $t \in T_1$ and an empty place p such that $p \in {}^\bullet t$, $^\bullet p \cap T_1 \neq \emptyset$. It follows from Definition 3.12, that the situation $t_1 \in {}^\bullet p$, $t_2 \in {}^\bullet p$, $t_1 \in T_1$, $t_2 \notin T_1$ is impossible. Then every transition $t' \in {}^\bullet p$ is in a block Σ' such that $(\Sigma', \Sigma_1) \in R$ and is disabled (because in such block there are no enabled transitions). To construct stubborn set T_S, we should add to T_1: transition t', all transitions from $^\bullet p'$, where p' is an empty place in $^\bullet t'$, and so on, while it is possible. Every transition t'' added in such way belongs to a block Σ'' such that $(\Sigma'', \Sigma_1) \in \widehat{R}$ and hence it is disabled. So, if T_1 is not a stubborn set, then there exists a stubborn set T_S such that $T_1 \subset T_S$ and $enabled(M) \cap T_S \setminus T_1 = \emptyset$. Then $enabled(M) \cap T_S = enabled(M) \cap T_1$ and, from Theorem 3.10, it is a persistent set.

The block decomposition method and the stubborn set method in their classical form have very similar purposes. So, two questions arise here: first, which of two approaches is preferable in a given situation; second, whether and how it is possible to combine them.

These questions can be answered in the following way: the stubborn set method is more universal, than the block decomposition method (it can be applied only to certain classes of nets, as it is described in Section 4.1); on the other hand, every block in this method is analyzed to detect the terminal states (deadlocks), so using the stubborn set method for analysis of blocks would be an evident improvement. However, there is one restriction: if the net should be checked for safeness (boundedness), generation of the full state space of a block cannot be substituted by generation of an RRG by means of the stubborn set method, because in some cases it would not allow to decide these properties.

What can be said about hierarchical decomposition and the persistent sets? Hierarchical decomposition can be applied to analysis of a wider range of properties than the classical stubborn set method; but any analysis based on the hierarchical decomposition, like the analysis based on the block decomposition, would require detection of terminal states of the blocks, which can be performed by means of the persistent set methods. Therefore the decomposition methods and the persistent set methods can be successfully combined (see also section "Discussion" in [214]).

4.4 Parallel Analysis

"Lazy state space construction" approaches for Petri net analysis use the fact that parallel branches of a Petri net can be considered as independent, so such approaches analyze the nets locally (the same is true for the reduction approach).

On the other hand, the algorithms of analysis usually remain sequential; and parallel structure of Petri nets would allow a essential reduction of analysis time, if the analysis is performed by a multiprocessor system. So the next problem arises and is considered in this section: a Petri net is given, and n parallel processes are given; each of them is able to perform certain operations. How the net can be efficiently analyzed by these processes?

Of course, the methods constructing the state spaces can be parallelized, and concurrent versions of the algorithmsalgorithm!parallel of lazy state space constructions, such as stubborn set method and the maximal concurrent simulation, can be developed. But every thread of such parallelized algorithmalgorithm!parallel would have to consider the whole net. So it seems to be more prospective to develop the parallel algorithmsalgorithm!parallel using decomposition methods.

Two questions should be answered for organizing efficient parallel analysis: how to distribute the functions between the processes and how to organize proper communication between them [130]. Some efforts have been made to parallelize examining protocol states (without direct use of Petri nets, see for example [203]), performing Petri net unfolding [87] and the complete reachability analysis (distributed algorithm for modular analyzer of Petri nets "Maria" [160]).

The block decomposition method can be parallelized in the natural way. Such parallelization is described and discussed below. In this section we suppose that the processes communicate through the common memory containing Petri net description and current data.

Let us try to produce the concurrent version of Algorithm 4.3. The problem is how to organize non-conflicting concurrent processing of the net and effective synchronization of the processes. Every process may analyze a block and then write the results of its work into the common memory. But in this approach there is no sense to use always the *minimal* blocks, because it may unnecessarily increase the synchronization costs.

The following algorithm of block decomposition is proposed to be used for parallel analysis. Let Σ be an OPN.

Algorithm 4.16

1. Find the minimal blocks by using transitive closure of the relation of alternative joint of transitions.
2. Unite all cycles composed by blocks.
3. Unite all complete sequential compositions of the blocks (see Fig. 4.2).
4. If some blocks are sufficiently small for evaluated time of their analysis be less than evaluated time of updating the set B^5 by a process, unite them with their neighbors. If some new complete sequential compositions have arisen, go to 3.
5. The end.

A concurrent version of Algorithm 4.3 is presented below.

---

[5] Where B is the set of markings, as in Algorithm 4.3.

Let Σ be an operational PN, M_0 its initial marking.

Algorithm 4.17

$f := 0$ (f enables/disables write access to the common data); $r := 0$ (a global variable containing error code).

1. Decompose the net Σ into blocks applying Algorithm 4.16.
2. Calculate the relation of partial order R.
3. $B := \{M_0\}$.
4. Parbegin. For every concurrent process
   a) Wait when there is a block Σ_i such that there is no block Σ_j not analyzed yet for which $(\Sigma_j, \Sigma_i) \in R$, and Σ_i is not marked as "under processing" or "analyzed", or there are no blocks not analyzed yet, or $r \neq 0$. If there are no blocks not analyzed or $r \neq 0$, go to 5.
   b) Mark Σ_i as being "under processing".
   c) For each marking of Σ_i being the projection of a marking $M \in B$, find the terminal markings of Σ_i reachable from it.
   d) If at least one of the initial markings for Σ_i is found to be incorrect, set $r := 1$ and go to 5.
   e) Wait until $f = 0$.
   f) Set $f := 1$.
   g) Replace markings in B, such that input places of Σ_i are marked by the markings, where the initial markings of Σ_i are replaced by correspondent terminal markings (tokens not belonging to the places of Σ_i do not change their positions).
   h) Set $f = 0$; mark Σ_i as "analyzed".
   i) Go to a).
5. Parend
6. If $r = 0$, the initial marking M_0 is correct, and B contains all terminal markings reachable from it in Σ. Else the initial marking is incorrect. The end.

Of course, this approach can be used also for analysis of a cyclic net satisfying condition, formulated in Subsection 4.1.3. The corresponding algorithm is presented in [130].

Let us try to evaluate effectiveness of parallelization of analysis algorithms. If the analysis of a net Σ by single process (using Algorithm 4.3) requires $c_1(\Sigma)$ time, how much time will be required for analysis of the same net by n processes ($c_n(\Sigma)$, Algorithm 4.17)?

First of all, it depends on the number of blocks that can be analyzed simultaneously. Sets of such blocks correspond to the cliques of the graph representing concurrency relation between the blocks. Let m be the average number of such blocks.

Then $c_n \approx (c_1 - c_i)/min(n, m) + c_i + c_s$, where c_i is the time needed for initialization, c_s is the time needed for synchronization between processes.

On the other hand, for effectiveness evaluation, the methods of critical path calculation in the project networks can be used [53, 171] - and the graph of partial order relation R between the blocks can be easily converted into a project network, if time of processing of every block is evaluated. Then the total time of processing by means of a multiprocessor system will correspond to the length of critical path in the network, and time of processing using single processor will be the sum of processing time needed for all blocks.

4.5 Distributed Analysis

4.5.1 A Method of Distributed Analysis

Algorithm 4.17 is intended for a multiprocessor structure with common memory. It can be implemented as a multithread application. But analysis of large nets would be more efficient by using computer networks, and in this case Algorithm 4.17 cannot be applied directly. In this section the modification intended for distributed analysis is described.

The problem with distributed version of the analysis algorithm is that although analysis of the blocks can be performed in parallel by different processors, at every step of analysis a set of markings of the whole net (set B in Algorithm 4.17) should be kept; in our method it is kept in a centralized way. So, we have chosen the following decision: there is a master computer which keeps description of set B and controls the analysis process. The rest of computers in the network are slaves; each slave has the description of some subnets (blocks) of the Petri net in its memory and, obtaining the sets of possible initial markings of the subnets from the master, performs the analysis. The master obtains the sets of reachable terminal markings of the subnets and the messages about incorrectness from the slaves .

Before starting the net analysis, the blocks of the decomposed net should be assigned to the computers in the way which minimizes the analysis time. According to the structure we have selected, there is no direct communication between slaves; the whole inter-computer communication goes through the master. We suppose that the number of input and output places of the blocks is much less than the number of their internal places (that is usual situation in practical applications), hence communication time is not taken into account in the algorithm of processor assignment (however, according to experimental results, it does affect the analysis time considerably and should be taken into account in further work). For optimal partitioning, the partial order in which blocks have to be analyzed and evaluated analysis time of each block should be considered.

We use the partitioning algorithm described in [164]. Our experiments with randomly generated nets show, that time complexity of a block analysis (by means of full state space construction) can be evaluated by the function $3^{(|P_i|/6)}$, where P_i is the set of places of the block. Using this evaluation, the program dependence graph (PDG, [164]) is built having the same structure as the oriented acyclic graph specifying relation R, and to every node i the weight w_i is assigned according to the following empirical formula:

$$w_i = 3^{(|P_i|/6)}. \tag{4.10}$$

The method of distributed analysis of OPNs is described below. It consists of 3 algorithms: initialization and the algorithms for master and for slave.

Let Σ be an OPN, M_0 its initial marking.

Algorithm 4.18

(Initialization)

1. Decompose the net Σ into minimal blocks.
2. Unite all cycles composed by the blocks.
3. Create a PDG: nodes correspond to the blocks; arcs correspond to relation R between the blocks. To every node a weight is assigned according to (4.10).
4. Build a parse tree for the PDG [164].
5. Perform processor assignment (using the algorithm from [164]).
6. Write to every processor's memory the descriptions of the blocks corresponding to the assigned nodes of PDG.

Algorithm 4.19

(Master)

1. $B := \{M_0\}$.
2. While not all blocks are analyzed, do:
   a) Find a block Σ_i such that Σ_i is not analyzed and there is no block Σ_j not analyzed yet, for which $(\Sigma_j, \Sigma_i) \in R$.
   b) For each marking $M \in B$ such that $M(P_i^{in}) > 0$, send the projection of M on Σ_i to the processor to which the block is assigned. Do not consider Σ_i as "not analyzed" any more.
   c) If, from any slave, the description of terminal markings of block Σ_i reachable from given initial markings is obtained, replace markings in B, such that input nodes of Σ_i are marked, by the markings, where the initial markings of Σ_i are replaced by the corresponding terminal markings.
   d) If, from any slave, the message is obtained that a marking is incorrect, go to 4.
3. The initial marking M_0 is correct and B contains all the terminal markings reachable from it in Σ. The end.
4. The initial marking is incorrect. The end.

Algorithm 4.20

(Slave)
When, from the master, description of an initial marking of block Σ_i (M_{i0}) is obtained:

1. Find all terminal markings reachable from M_{i0}.
2. If M_{i0} is correct for Σ_i, send to master description of all terminal markings reachable from it. Else send the message that M_{i0} is incorrect for Σ_i.

4.5.2 Implementation of the Method

The method described above has been implemented in Java[6]. This programming language has been chosen for the following reasons: availability of the mechanisms supporting distributed programming, support of multithreading and availability of parsers for XML language. As a distributed environment for the project, Java Parallel Virtual Machine (JPVM) [69] was selected. Advantages of JPVM are: its interface close to the interface of Parallel Virtual Machine (PVM) [72] and semantics fitting for the object-oriented programming language Java. JPVM has been completely implemented in Java, and can be used in a heterogeneous distributed environment.

Input of the program implementing the method is an OPN described in the format PNSF3 [?]. The input data are initially validated according to the corresponding document type definition (DTD). The master performs block decomposition of the net. After that, according to the obtained block structure, the master (Fig. 4.10) performs processor assignment by means of Algorithm 4.18. The threads are initiated to control the slaves. Number of threads corresponds to the number of blocks ready to be processed, and number of the initiated threads corresponds to the number of slaves, each of which has some blocks to process, assigned by the module ProcessorAssignment. At each step a thread sends the initial markings of the corresponding block(s) to the slave. A slave's task is verification of the given block and computing of its terminal markings. Result of the slave's work is sent back to the master, to the corresponding thread being in waiting mode. If the initial markings are correct for the block, the description of its terminal markings is sent to the main process of the master for further analysis of the net; otherwise the program exits and further analysis is not performed.

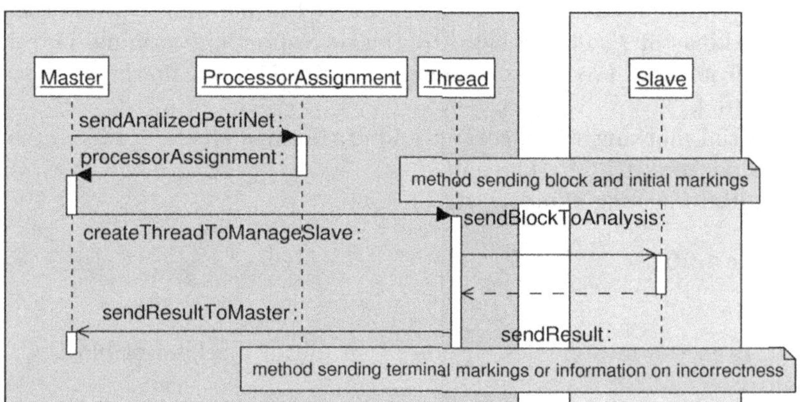

Fig. 4.10. Distributed analysis of Petri nets: operating model of Master and Slave

---

[6] The method was implemented by Tomasz Gratkowski [128].

Table 4.3. Time of the distributed analysis of Petri nets

computers	net parameters		
	128×118	238×228	360×349
1	2377	7277	12095
3	3202	6684	9092
5	3060	5893	8920
7	2626	5208	7578

JPVM environment serves as an intermediate layer between the master and the slaves. The master sends the messages to the slaves by its threads, ordering to the JPVM demons of the slaves to initialize the appropriate analysis procedures. In the next JPVM messages, input data for these procedures are sent. The result of slave's work is a JPVM message with the description of terminal markings or a message describing errors.

4.5.3 Experimental Results and Concluding Remarks

For experimental analysis of the approach presented above, some tests have been performed. The system was tested at a separated local-area network based on 100Base-T technology. The network consisted of 7 computers at most. Parameters of the analyzed nets and the analysis time (in ms) are given in Table 4.3.

The obtained results are modest, but they show that the method seems to be promising for large nets. For small nets, communication between the computers within the network noticeably decreases the gain (as it can be seen in Table 4.3, for the smallest of considered nets there is no gain at all). For larger nets the method provides greater possibilities of time saving. This correlates with results of the experiments on the sequential version of analysis based on block decomposition (see [251] and subsection 4.1.4).

Several problems are to be solved here. One of them is the evaluation of analysis time of a single block; formula (4.10) is far from being exact. Another problem is to take into account the costs of data transfer in the network, important for analysis of some kinds of nets. Last bat not least, optimization of data transfer - there are grounds to suppose, that data transfer in our system can be noticeably improved.

5. Analysis by Solving Logical Equations – Calculation of Siphons and Traps

This chapter is dedicated to one of the important Petri net analysis tasks - calculation of siphons and traps. For some classes of Petri nets (and their extensions) properties like liveness and reversibility can be decided by analysis of siphons and traps [21, 28, 155, 176, 194, 228, 241, 249] - for example, there is a known result that a free choice net is live, if and only if every siphon contains a marked trap [140, 183]. Finding siphons and traps has a variety of applications to verification and design of parallel systems, such as detection and prevention of deadlocks (usually in a correct system no siphon can be emptied) [65, 81, 92]; one application of this kind will be presented below in Subsection 6.1.3.

Siphons and traps of a Petri net correspond to the roots of certain logical equations, which can be represented in CNF. Thelen's method [162, 207] allows efficient calculation of prime implicants of a Boolean function represented in such form. Application of the Thelen's method for solving the mentioned logical equations is discussed below; it makes possible obtaining sets of siphons and traps in form of ternary vectors. Some heuristics for Thelen's method have been proposed and the computer experiments performed.

Normal Forms and Implicants of Boolean Expressions

Here we recall some definitions from Boolean algebra, which will be necessary below.

Definition 5.1. *A literal is either a propositional letter x or negation $\bar{x}$ of a propositional letter x.*

Definition 5.2. *A conjunctive normal form (CNF) formula is one which is a conjunction of disjunctions of literals. That is, a formula of the form:*

$$(l_{11} \vee \ldots \vee l_{1n_1}) \wedge \ldots \wedge (l_{m1} \vee \ldots \vee l_{mn_m}), \tag{5.1}$$

where each l_{ij} is a literal.

A. Karatkevich: Dynamic Analysis of Petri Net-Based Discrete Sys., LNCIS 356, pp. 87–93, 2007.
springerlink.com

Definition 5.3. *A disjunctive normal form (DNF) formula is one which is a disjunction of conjunctions of literals. That is, a formula of the form:*

$$(l_{11} \wedge \ldots \wedge l_{1n_1}) \vee \ldots \vee (l_{m1} \wedge \ldots \wedge l_{mn_m}), \tag{5.2}$$

where each l_{ij} is a literal.

Definition 5.4. *An implicant k of a Boolean function f is a conjunction of literals such that $k \to f$. A prime implicant k_p of a Boolean function f is such an implicant of f that removing any literal from k_p leads to a conjunction k' such that k' is not an implicant of f.*

5.1 Known Methods of Calculation of Siphons and Traps

5.1.1 Calculation of Siphons and Traps by Means of Solving Logical Equations

Siphons and traps do not depend on the initial markings and can be detected by structural analysis of a net. A siphon D has to satisfy the set of conditions: $\forall t_i \in T: t_i^{\bullet} \cap D \neq \emptyset: \Rightarrow {}^{\bullet}t_i \cap D \neq \emptyset$ (see subsection 2.1.2).

Analogous conditions exist for a trap. They can be in natural way represented by logical equations.

Consider a set of Boolean variables $\mathbb{P} = \{\mathbf{p}_1, \mathbf{p}_2, ..., \mathbf{p}_m\}$, where $m = |P|$, and $\mathbf{p}_i = 1$ if and only if p_i belongs to a certain subset of places. Denote the subsets of $\mathbb{P}$ corresponding to ${}^{\bullet}t_i$ and $t_i^{\bullet}$ by $\mathbb{P}_i$ and $\mathbb{P}_i^*$. Denote disjunctions of the corresponding variables by $\vee[\mathbb{P}_i]$ and $\vee[\mathbb{P}_i^*]$. Then the condition for transition t_i can be expressed by the formula: $\vee[\mathbb{P}_i^*] \to \vee[\mathbb{P}_i] = 1$.

Then all the siphons of Petri net $\Sigma = (P, T, F)$ are specified by the roots of the logical equation.

Lemma 5.5 (Affirmation 4.9 from [249]). *All siphons of a Petri net Σ are defined by the roots of logical equation*

$$\wedge_{i=1}^{n}(\vee[\mathbb{P}_i^*] \to \vee[\mathbb{P}_i]) = 1, \tag{5.3}$$

where $n = |T|$.

The analogous lemma can be formulated for traps.

Lemma 5.6 (Affirmation 4.8 from [249]). *All traps of a Petri net Σ are defined by the roots of logical equation*

$$\wedge_{i=1}^{n}(\vee[\mathbb{P}_i] \to \vee[\mathbb{P}_i^*]) = 1, \tag{5.4}$$

where $n = |T|$.

So, siphons and traps of a Petri net can be calculated by solving logical equations; this methodology is described in [249]. How to solve efficiently the equations of this kind? One of the ways is elimination of implications and transformation of the equation into DNF, from which the roots can be obtained. Another approach allows simplifying the calculations, taking into account that in (5.3) and (5.4) all variables occur without negation. In that approach the ternary matrices are used to represent the equations, and the roots are obtained by combinatorial operations on those matrices. For details see [249]. Another approach to symbolic calculation of siphons and traps is described in [192, 193].

Let us consider some transformations which can simplify solving the equations. Let $m = |t_i^\bullet|$, $\mathbf{p}_1^{i^*} ... \mathbf{p}_m^{i^*}$ - variables corresponding to the output places of t_i. Then:

$$\vee[\mathbb{P}_i^*] \to \vee[\mathbb{P}_i] = \overline{\vee[\mathbb{P}_i^*]} \vee (\vee[\mathbb{P}_i]) = \wedge_{j=1}^m \left(\overline{\mathbf{p}_j^{i^*}} \vee (\vee[\mathbb{P}_i]) \right). \tag{5.5}$$

The equation is now transformed into CNF. Note that changing interpretation of the variables, such that belonging of a place to a set means that the corresponding variable has value 0, would lead to inversion of all the literals in (5.5), and the expression turns to be a conjunction of Horn clauses. As far as we know, Minoux and Barkaoui [170] were the first who considered siphons and traps in Petri nets as Horn-satisfiability solutions.

There is an efficient polynomial-time algorithm solving the satisfability problem for the Horn expression and finding a decision with minimal number of variables having value 1 [178]. Hence, for a given subset of places it is possible to calculate efficiently *the maximal siphon contained in it*; this approach is used in [21]. There are also methods of finding the minimal siphons or the minimal siphons containing given places [206, 230, 231]. But here the task of generating all siphons is considered; applying the mentioned approach to this task would require a time-consuming combinatorial search.

Note that the approach described in this subsection does not lead to finding basic siphons (traps) only, unlike the approaches described below. But this does not mean, that all the siphons (traps) are calculated and represented explicitly; it would lead to non-compact representations. A system of Boolean equations of the kind (5.3) or (5.4) can be solved in such a way that the set of its roots will be represented by a ternary matrix, every row of which describes several solutions (see an example below). That is how we are going to solve the task.

5.1.2 Other Approaches to Calculation of Siphons and Traps

There are several methods in which siphons and traps are calculated by solving systems of linear equations or inequalities. These methods can be grouped into 3 cathegories [66].

1. The particular class of Boolean equation systems can be transformed into systems of linear inequalities, by solving which the siphons or traps can be found [198].
2. The so-called *p-invariants* can be obtained by solving a linear equation $Cx = 0$, where C is the incidence matrix of a net [172]. The set of states corresponding to a *p*-invariant is an *st*-component (a siphon and a trap at the

same time). But not every siphon and trap has a corresponding p-invariant. In the second approach [155] the net is transformed in such a way that for every siphon (or trap) of the initial net there is corresponding p-invariant of the transformed net. So, calculation of the siphons (traps) of the initial net can be performed by solving linear equations generated according to the transformed net.

3. A support of a vector being a solution of linear inequality $Cx \geqslant 0$ is a trap in the corresponding net; but not every trap corresponds to such a solution. In the third approach [197] the incidence matrix is transformed in certain way so that it describes the same graph of a net, but with weighted arcs. Solving the inequality obtained in this way allows to detect the traps of the net. As far as the tasks of calculation of siphons and traps are equivalent (a siphon of a net Σ is a trap in the net obtained from Σ by reversing all arcs), the approach can be used for calculation of siphons too.

This chapter is dedicated to calculation of siphons and traps by solving logical equations, so the methods using linear algebra are not considered here in details.

5.2   Algorithm to Find Siphons and Traps

As it is shown in Subsection 5.1.1, a Boolean formula describing siphons or traps of a Petri net can be easily transformed into CNF (5.5). So the Thelen's prime implicant method can be efficiently applied to calculating of siphons and traps. It was suggested in [229] to apply the Thelen's method for Petri net analysis. The new contribution is the efficient heuristics for finding prime implicants. These heuristics (reordering the sum terms and the literals in the sum terms) can be applied not only for Petri net analysis, but also for solving other tasks where the Thelen's method is used [32, 33, 228]. Thelen's prime implicant method, the heuristics and the experimental results are described in details in Appendix D. The algorithm of calculating siphons and traps is presented below.

Let Σ be a Petri net.

Algorithm 5.7

1. Generate the Boolean formula for siphons (traps) of the Petri net according to equations (5.3), (5.4).
2. Transform the formula into CNF applying equation (5.5).
3. Sort the sum terms and the literals in the sum terms to minimize search tree.

    a) Sort the sum terms by variables (Heuristic 2a from Appendix D).

    b) Sort the literals in the sum terms (Heuristic 4 from Appendix D).

4. Find all prime implicants of the obtained formula using Thelen's method.
5. Represent set of prime implicants as a ternary matrix. Each row corresponds to one prime implicant and represents one or more siphons (traps).

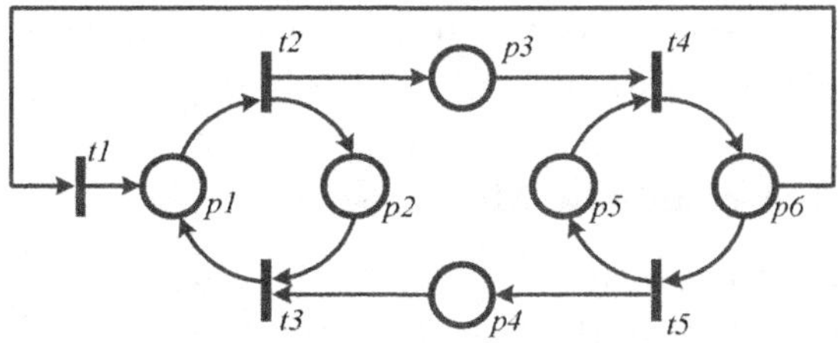

Fig. 5.1. A Petri net [241]

5.3   Example

Calculation of siphons and traps of a sample Petri net by using the proposed method and some other methods is shown below.

5.3.1   The Proposed Method

Let us find all siphons of the Petri net from Fig. 5.1. First, Boolean formula is generated according to equation (5.3):

$$(\mathbf{p_1} \rightarrow \mathbf{p_6})(\mathbf{p_2} \vee \mathbf{p_3} \rightarrow \mathbf{p_1})(\mathbf{p_1} \rightarrow \mathbf{p_2} \vee \mathbf{p_4})$$
$$(\mathbf{p_6} \rightarrow \mathbf{p_3} \vee \mathbf{p_5})(\mathbf{p_4} \vee \mathbf{p_5} \rightarrow \mathbf{p_6}) = 1. \tag{5.6}$$

Next the formula is transformed into CNF using equation (5.5):

$$(\overline{\mathbf{p}}_1 \vee \mathbf{p_6})(\overline{\mathbf{p}}_2 \vee \mathbf{p_1})(\overline{\mathbf{p}}_3 \vee \mathbf{p_1})(\overline{\mathbf{p}}_1 \vee \mathbf{p_2} \vee \mathbf{p_4})$$
$$(\overline{\mathbf{p}}_6 \vee \mathbf{p_3} \vee \mathbf{p_5})(\overline{\mathbf{p}}_4 \vee \mathbf{p_6})(\overline{\mathbf{p}}_5 \vee \mathbf{p_6}) = 1. \tag{5.7}$$

All prime implicants are found and the tree size for different heuristics (described in Appendix D) is shown below:

No heuristics – order like in (5.7) 37 nodes,

Sort by Length   27 nodes,

Sort by Variables   25 nodes,

Reordering Literals   37 nodes,

SV + RL   25 nodes.

Prime implicants are represented by a ternary matrix:

$$
\begin{array}{cccccc}
p_1 & p_2 & p_3 & p_4 & p_5 & p_6
\end{array}
$$

$$
\begin{bmatrix}
0 & 0 & 0 & - & 1 & 1 \\
- & 0 & 0 & 1 & 1 & 1 \\
1 & 1 & 1 & - & - & 1 \\
1 & 1 & - & - & 1 & 1 \\
1 & - & 1 & 1 & - & 1 \\
1 & - & - & 1 & 1 & 1
\end{bmatrix} . \tag{5.8}
$$

For the analyzed Petri net, there are 11 siphons, e.g. the first row defines 2 siphons: $\{p_5, p_6\}$ and $\{p_4, p_5, p_6\}$.

5.3.2 Some Other Symbolic Methods

By using the method described in [249], the task would be solved as follows. Expression in the left part of (5.6) is transformed into a product of DNFs and then into DNF:

$$(\overline{\mathbf{P}}_1 \vee \mathbf{P}_6)(\overline{\mathbf{P}}_2\overline{\mathbf{P}}_3 \vee \mathbf{P}_1)(\overline{\mathbf{P}}_1 \vee \mathbf{P}_2 \vee \mathbf{P}_4)(\overline{\mathbf{P}}_6 \vee \mathbf{P}_3 \vee \mathbf{P}_5)(\overline{\mathbf{P}}_4\overline{\mathbf{P}}_5 \vee \mathbf{P}_6)$$
$$= \overline{\mathbf{P}}_1\overline{\mathbf{P}}_2\overline{\mathbf{P}}_3\overline{\mathbf{P}}_4\overline{\mathbf{P}}_5\overline{\mathbf{P}}_6 \vee \overline{\mathbf{P}}_1\overline{\mathbf{P}}_2\overline{\mathbf{P}}_3\mathbf{P}_5\mathbf{P}_6 \vee \overline{\mathbf{P}}_2\overline{\mathbf{P}}_3\mathbf{P}_4\mathbf{P}_5\mathbf{P}_6$$
$$\vee \mathbf{P}_1\mathbf{P}_2\mathbf{P}_3\mathbf{P}_6 \vee \mathbf{P}_1\mathbf{P}_2\mathbf{P}_5\mathbf{P}_6 \vee \mathbf{P}_1\mathbf{P}_3\mathbf{P}_4\mathbf{P}_6 \vee \mathbf{P}_1\mathbf{P}_4\mathbf{P}_5\mathbf{P}_6 = 1. \tag{5.9}$$

The sum of products represents the set of siphons similarly as this set is represented by (5.8). Transformation (5.9) is performed by direct multiplying of the expressions in parentheses; Thelen's method cannot be used here because the sums to be multiplied are not elementary. For this example, such transformation requires 31 multiplications of the elementary sum terms (compare with 24 multiplications of an elementary sum term and a literal when our method is used).

In another approach [241,249] a ternary matrix W_r is constructed corresponding to (5.3); for this example it looks as follows:

$$
\begin{array}{c}
p_1\,p_2\,p_3\,p_4\,p_5\,p_6 \\
W_r = \begin{bmatrix}
0 & - & - & \sim & - & 1 \\
1 & 0 & 0 & \sim & - & - \\
0 & 1 & - & 1 & - & - \\
- & - & 1 & \sim & 1 & 0 \\
- & - & - & 0 & 0 & 1
\end{bmatrix}.
\end{array}
\tag{5.10}
$$

In this matrix the column minors not containing the so-called negative rows (rows with 0-elements, but without 1-elements) correspond to the siphons. To find them, the combinatorial search is necessary. For example, two last columns satisfy the property just mentioned and correspond to the siphon $\{p_5, p_6\}$.

5.3.3 The Linear Algebraic Method

The linear algebraic method [66] requires to solve the system of linear inequations $y^T C_\Theta \geqslant \mathbf{0}$, where y is a vector indexed by places of a net, C_Θ is a transformed incidence matrix which, for our example, may look as follows:

$$
\begin{array}{c}
\quad\ \ t_1\ \ t_2\ \ t_3\ \ t_4\ \ t_5 \\
\begin{array}{c}
p_1 \\ p_2 \\ p_3 \\ p_4 \\ p_5 \\ p_6
\end{array}
\left[
\begin{array}{rrrrr}
-1 & 2 & -1 & 0 & 0 \\
0 & -1 & 1 & 0 & 0 \\
0 & -1 & 0 & 1 & 0 \\
0 & 0 & 1 & 0 & -1 \\
0 & 0 & 0 & 1 & -1 \\
1 & 0 & 0 & -1 & 2
\end{array}
\right].
\end{array}
\qquad (5.11)
$$

Support of any integer non-negative solution of the system corresponds to a siphon. For example, $(0, 0, 0, 0, 1, 1)$ is a solution, and it corresponds to the siphon $\{p_5, p_6\}$.

5.4 Concluding Remarks

The first two methods of four demonstrated in Section 5.3 differ in the way of transformation of equations; the first one includes the optimization techniques and allows to obtain the result in smaller number of elementary steps (multiplications). Two last methods lead to obtaining the sets of siphons (traps) in explicit representation, and two first approaches represent these sets in compact form of ternary matrices. Compact representation is certainly a remarkable advantage here, because the number of siphons and traps is exponential in the size of the net. Compactness of representation of the sets of traps (siphons) can be increased by using the methods of minimization of Boolean functions [166]. Another prospective way of obtaining compact descriptions of siphons and traps can be based on binary decision diagrams (BDD); from a BDD representation of the functions in left parts of equations (5.3) and (5.4) it would be easy to get information on siphons and traps. This requires however additional research on the subject.

6. Verification of Detailed System Descriptions

6.1 Application of the Described Approaches to Other Parallel Discrete Models

6.1.1 Interpreted Petri Nets and Sequent Automata

Analysis of Sequent Automata and Nets with Internal Variables

Reachability graph G_{int} of an interpreted Petri net Σ_{int} is a subgraph of the reachability graph G of the underlying net Σ, because an interpretation can reduce possibilities of the net evolution and never can expand them. RRG G_R, constructed for Σ, is a subgraph of G, but not necessarily of G_{int}; so G_R can miss some information, important for analysis of Σ_{int}.

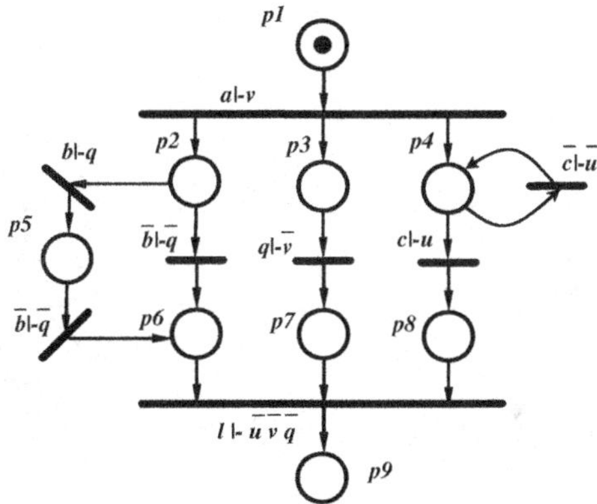

Fig. 6.1. An interpreted Petri net

Consider the following example. Petri net from Fig. 3.1 has one deadlock, which can be detected by the stubborn set method. The interpreted Petri net shown in Fig. 6.1, with the same underlying net, has two reachable deadlocks - if

A. Karatkevich: Dynamic Analysis of Petri Net-Based Discrete Sys., LNCIS 356, pp. 95–121, 2007.
springerlink.com

the initial value of q is 0, and if $b = 0$ when place p_2 has a token, then a token cannot leave place p_3, because the condition $q = 1$ of firing of t_5 is never satisfied. So, marking at which places p_3, p_6, p_8 have tokens, may be a deadlock.

The stubborn set method, applied to the underlying net, cannot detect it. If during the RRG constructing the internal variables are taken into account, but the stubborn sets are constructed according to the basic method, the results may be interesting, but still unsatisfactory.

Applying the stubborn set method to a non-interpreted Petri net, we often have more than one variant of stubborn set for a given marking. Size of RRG may remarkably depend on those choices, but in any case the set of detected deadlocks will be the same [215]. If we apply the method to an interpreted net, the deadlocks may be detected or not, dependently on the selected stubborn sets. For our example, the variants of RRG are shown in Fig. 6.2. The first variant shows that two deadlocks are possible. However, the second variant finds only one deadlock, the other one remains undetected.

This example shows, that the stubborn set method in its classical form does not work properly for interpreted Petri nets. The notion of stubborn set has to be re-defined for such nets to take into account interaction via internal variables. And an algorithm allowing deadlock detection in interpreted Petri nets would work also for sequent automata, because a sequent automaton can be described as a form of interpreted Petri net (and vice versa). So, it makes sense to concentrate on deadlock detection for sequent automata; in this way a deadlock detection method for the interpreted Petri nets will be obtained too.

Let us start form defining the notion of independent sequents, being a concretization of the notion of independent transitions (Definition 3.1). For the

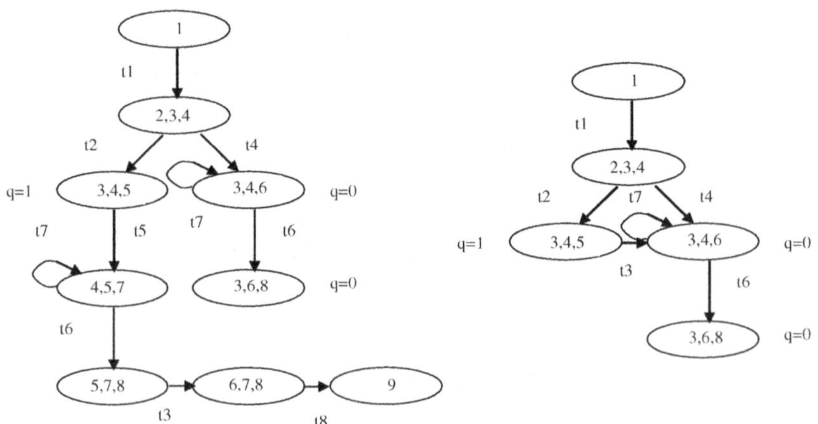

Fig. 6.2. The variants of RRG for the nets from Fig. 6.1

sequents (and the transitions of the interpreted Petri nets), unlike for the transitions of the classical Petri nets, the second condition is not redundant, because the diamond rule does not hold.

Definition 6.1. *Sequents $\varphi_1 \vdash \psi_1$, $\varphi_2 \vdash \psi_2$ are (globally) independent, if the following conditions hold:*

1. *no variable appearing in ψ_1 (ψ_2) is an argument of φ_2 (φ_1) (independent sequents can neither disable nor enable each other);*
2. *$\psi_1\psi_2 \not\equiv 0$ (commutativity of enabled independent sequents).*

Definition 3.16a can be concretized for the sequent automata. In order to do it, we have to describe the case when a sequent can enable another one and when it can disable another one (see Section 3.2). It can be done by obtaining CNF and DNF forms for left part of every sequent, as it is described in [117]; but there is a simpler way, based on transformation of a general-case sequent automaton into equivalent simple sequent automaton. This can be done for every sequent automaton by transforming left part φ_i of every sequent s_i into DNF (in the applications it should be minimized) and splitting every sequent into several simple sequents, such that every product term of φ_i will be the left part of a sequent, and ψ_i will be its right part [249].

If all the sequents are simple, then s_1 can enable s_2, if there is a literal l_i occurring in ψ_1 and φ_2, and in the current state $l_i = 0$; s_1 can disable s_2, if there is variable x, occurring in ψ_1 with (without) negation and in φ_2 without (with) negation. Every literal occurring in the left part of a sequent specifies a necessary condition of its enabling.

Definition 6.2. *A subset S_S of the sequents of a simple sequent automaton at state M is a stubborn set, if (1) for every sequent in S_S such that there is a literal $l = 0$ in its left part, being an internal variable with or without negation, every sequent such that l occurs in its right part belongs to S_S; (2) for every enabled[1] sequent s_i in S_S every sequent s_j such that $\psi_i\varphi_j \equiv 0$ or $\psi_j\varphi_i \equiv 0$ belongs to S_S; (3) for every enabled sequent s_i in S_S every sequent s_j such that $\psi_i\psi_j \equiv 0$ belongs to S_S; (4) S_S contains an enabled sequent s_i such that $\psi_i = 0$ at M.*

It follows from Theorem 3.18, that a reduced reachability graph of a simple sequent automaton built using the stubborn sets in the sense of the definition above contains all reachable deadlocks of the automaton. A safe Petri net, interpreted or not, can be easily described by a system of sequents [117]. Sequent automaton (6.1) corresponds to the net from Fig. 3.1, (6.2) - to the net from Fig. 6.1. Variables $\mathbf{p}_i$ correspond to places p_i.

---

[1] Or can become enabled for some combination of values of input variables; below the term "enabled" is used in this extended sense.

$$
\begin{aligned}
\mathbf{p}_1 &\vdash \overline{\mathbf{p}_1}\mathbf{p}_2\mathbf{p}_3\mathbf{p}_4 \\
\mathbf{p}_2 &\vdash \overline{\mathbf{p}_2}\mathbf{p}_5 \\
\mathbf{p}_2 &\vdash \overline{\mathbf{p}_2}\mathbf{p}_6 \\
\mathbf{p}_3 &\vdash \overline{\mathbf{p}_3}\mathbf{p}_7 \\
\mathbf{p}_4 &\vdash \overline{\mathbf{p}_4}\mathbf{p}_8 \\
\mathbf{p}_5 &\vdash \overline{\mathbf{p}_5}\mathbf{p}_6 \\
\mathbf{p}_6\mathbf{p}_7\mathbf{p}_8 &\vdash \overline{\mathbf{p}_6}\overline{\mathbf{p}_7}\overline{\mathbf{p}_8}\mathbf{p}_1
\end{aligned}
\tag{6.1}
$$

$$
\begin{aligned}
\mathbf{p}_1 a &\vdash \overline{\mathbf{p}_1}\mathbf{p}_2\mathbf{p}_3\mathbf{p}_4 v \\
\mathbf{p}_2 b &\vdash \overline{\mathbf{p}_2}\mathbf{p}_5 q \\
\mathbf{p}_2 \bar{b} &\vdash \overline{\mathbf{p}_2}\mathbf{p}_6 \bar{q} \\
\mathbf{p}_3 q &\vdash \overline{\mathbf{p}_3}\mathbf{p}_7 \bar{v} \\
\mathbf{p}_4 c &\vdash \overline{\mathbf{p}_4}\mathbf{p}_8 u \\
\mathbf{p}_4 \bar{c} &\vdash \bar{u} \\
\mathbf{p}_5 \bar{b} &\vdash \overline{\mathbf{p}_5}\mathbf{p}_6 \bar{q} \\
\mathbf{p}_6\mathbf{p}_7\mathbf{p}_8 &\vdash \overline{\mathbf{p}_6}\overline{\mathbf{p}_7}\overline{\mathbf{p}_8}\mathbf{p}_1 \bar{u}\bar{v}\bar{q}
\end{aligned}
\tag{6.2}
$$

We propose the following algorithm of deadlock detection for the interpreted Petri nets, based on the reasoning presented above. The algorithm constructs a reduced reachability graph for given safe interpreted Petri net Σ_{int}.

Algorithm 6.3

1. Obtain the sequent automaton equivalent to the net Σ_{int}.
2. If the sequent automaton is not simple, transform it into an equivalent simple sequent automaton.
3. Remove all input variables from the sequents.
4. Generate a reduced reachability graph of the automaton in the following way. For every state under consideration, simulate only the firing of those sequents, which are enabled and belong to the stubborn set S_S satisfying Definition 6.2.

Let us apply the method to the interpreted Petri net from Fig. 6.1. Fig. 6.3 shows the RRG for this net (in the nodes, internal and output variables having value 1 in the corresponding states are listed). It is one of several possible RRGs, but every RRG created with the algorithm described above contains the same deadlocks.

Of course, to detect the deadlocks in a sequent automaton (simple sequent automaton), Algorithm 6.3 without the first step (first two steps) can be used.

Analysis of the nets with inhibitor arcs

See the net of Fig. 6.4a. The inhibitor arc (p_5, t_1) may cause a deadlock: $\{p_1, p_5, p_6\}$. An RRG of the net shown in Fig. 6.4b does not contain this deadlock. This simple example demonstrates, that the basic stubborn set method has to be modified to work properly for the nets with inhibitor arcs.

In order to apply the general approach described in Section 3.2, we have to find out, which additional cases of enabling and disabling of transition firing exist in the nets with inhibitor arcs, in comparison with classical Petri nets. There are two such cases:

- transition t can disable transition t', if there is place $p \in t^{\bullet}$ such that $p \in (^{\bullet}t')^{(inh)}$;

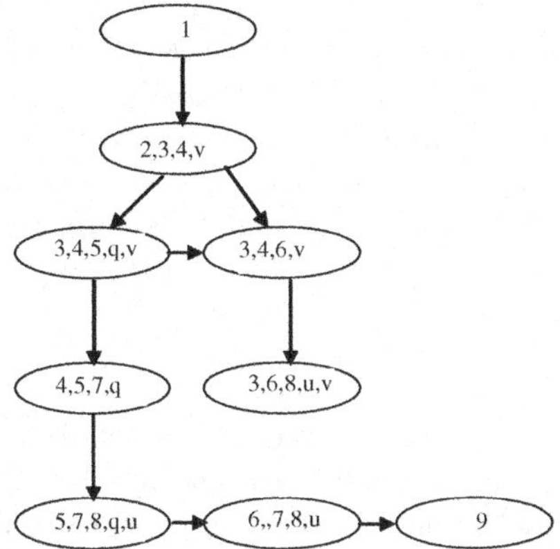

Fig. 6.3. RRG for the interpreted Petri net from Fig. 6.1, built by Algorithm 6.3

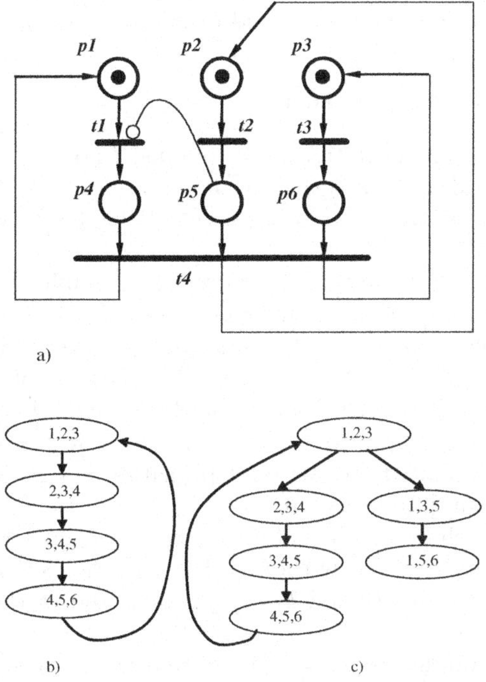

Fig. 6.4. A net with an inhibitor arc and two reduced reachability graphs

- transition t can enable transition t', if there is place $p \in {}^{\bullet}t$ such that $p \in ({}^{\bullet}t')^{(inh)}$.

Taking into account these two cases, the Definition 3.16 can be reformulated for the Petri nets with inhibitor arks as follows.

Definition 6.4. *A set T_S of the transitions of a Petri net with inhibitor arcs at marking M is a stubborn set, if (1) every disabled transition t in T_S has an empty input place p such that all transitions in ${}^\bullet p$ are in T_S or a marked input place p, such that $p \in ({}^\bullet t)^{(inh)}$ and all transitions in ${}^\bullet p$ are in T_S; (2) no enabled transition in T_S has a common input place with any transition (including disabled ones) outside T_S; (3) if there is an enabled transition t in T_S and place p such that $p \in t^\bullet$ and $p \in ({}^\bullet t')^{(inh)}$, then $t' \in T_S$; (4) if there is an enabled transition t in T_S and place p such that $p \in ({}^\bullet t)^{(inh)}$, then all transitions in ${}^\bullet p$ are in T_S; (5) T_S contains an enabled transition.*

It follows from Theorem 3.18, that a reduced reachability graph of a Petri net with inhibitor arcs built using the stubborn sets in sense of Definition 6.4 contains all reachable deadlocks of the net. Corresponding graph for the net of Fig. 6.4a is shown in Fig. 6.4c.

Sometimes, in the extended Petri net models for control systems specification, the *enabling arcs* are used [11, 12, 43]; defining the stubborn sets for such nets is easy, because they can be easily modelled by classical Petri nets. The corresponding definition is presented and discussed in [125].

Analysis of Nets with Priorities

In [30] a method of transformation of a bounded Petri net with priorities into a behaviorally equivalent classical Petri net is described. This equivalence is not complete, but preserves certain important behavioral properties, including reachable deadlocks.

The idea of the method can be briefly described as follows: for any transition t such that $\exists t' : (t', t) \in \rho$ an additional place p_t is introduced, which is an input and output place for t' and is connected to other transitions of the net in such way, that $M(p_t) + M({}^\bullet t) = const$. Then p_t is empty, if and only if t is enabled, which makes firing of t' impossible. Unfortunately, a safe ordinary net with priorities can be transformed by this method into an unsafe (but always bounded) net with weighted arcs without priorities. For details see [30]. Below, an example taken from [30] is shown.

See Fig. 6.5. Note that the reachability graph of the net with priorities (Fig. 6.5a) and of the equivalent net without priorities (Fig. 6.5c) are the same, but in the first case transitions t_1 and t_3 are parallel, and in the second case they are in conflict.

As in previous examples, we can see here, that the basic stubborn set method does not detect all deadlocks; for the initial marking of the net shown in Fig. 6.5a $\{t_3\}$ is a stubborn set, but if only the firing of transition t_3 is executed from the initial marking, the deadlock $\{p_2, p_5\}$ will never be reached.

As far as there exists a method of transformation of a net with priorities into a net without priorities, which can be successfully analyzed by means of the stubborn set method, it is possible to define the stubborn sets for the nets with priorities based on this transformation. Such definition looks as follows.

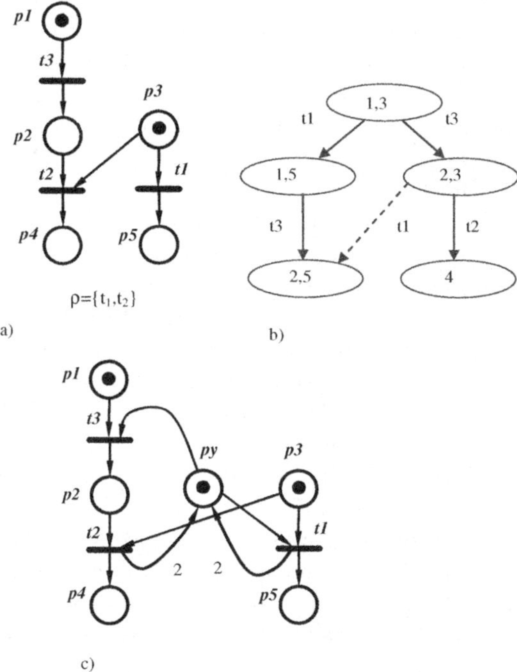

Fig. 6.5. A Petri net with priorities (a), its reachability graph (b) (the priority makes impossible the change of marking corresponding to the dashed arc), and the equivalent net without priorities (c) (thickened are the weighted arcs; p_y is the additional place)

Definition 6.5. *A set T_S of transitions of a Petri net with static priorities at marking M is a stubborn set, if: (1) every disabled transition in T_S has an empty input place p such that all transitions in $^\bullet p$ are in T_S; (2) no enabled transition in T_S has a common input place with any transition (including disabled ones) outside T_S; (3) T_S contains an enabled transition; (4) if an active transition in T_S cannot fire because there are enabled transitions with higher priority, then one of such transitions is in T_S; (5) if an enabled transition in T_S has an output place being an input place for a transition t such that $(t', t) \in \rho$, then $t' \in T_S$; (6) if for enabled transition $t \in T_S$ $(t, t') \in \rho$, then each transition, which input place is an output place of t', is in T_S.*

It follows from Theorem 3.9 and correctness of the Best-Koutny's method of transformation of a net with priorities into a net without priorities (proof see in [30]), that a reduced reachability graph of a Petri net with priorities built using the stubborn sets in sense of Definition 6.5, contains all reachable deadlocks of the net. For the example shown in Fig. 6.5, the graph built in such way is identical to the full reachability graph (at the initial marking the only set of transitions satisfying Definition 6.5 is $\{t_1, t_3\}$, so both deadlocks will be detected). Some other examples will follow.

However, condition (6) of Definition 6.5 is redundant. It takes into account, that if firing of an active transition t adds tokens to the input places of a transition t' such that $(t'', t') \in \rho$, then such firing can make impossible firing of t''. However, if t' and t'' have no common input places, this situation is temporal (before firing t') and does not affect the reachable deadlocks; and if t' and t'' share an input place, then t' is in T_S due to condition (2). That's why condition (6) can be removed from Definition 6.5 [132].

A net with priorities shown In Fig. 6.6a has one reachable deadlock less than the underlying net without priorities. Fig. 6.6b demonstrates its reduced reachability graph, built by means of the presented method. In Fig. 6.6c, a net with priorities is shown (a modified example from [27]), for which the basic stubborn set method can miss two reachable deadlocks - $\{p_2, p_6\}$ and $\{p_3, p_6\}$ (firing of transition t_1 would not be simulated at the initial marking - the dashed arc in Fig. 6.6d). The method proposed above allows detecting all the deadlocks.

It is interesting to compare the results presented above with [222], solving the same problem and also referring to [30], however with another structure of the proofs. The stubborn sets for priority nets described by Varpaaniemi allow better reduction of reachability graphs, than the sets satisfying Definition 6.5, but they are more complicated in their definition and more difficult to construct. We can say, that in [222] a generalization of weak stubborn sets for the nets with priorities is proposed, and Definition 6.5 presents a similar generalization of strong stubborn sets.

6.1.2   Statecharts

Static Analysis

Possible deadlocks in Statecharts can be detected by solving a system of Boolean equations, such that its roots specify the deadlocks.

As far as in a deadlock, according to definition, no transition can fire, for every active local deadlocked state p the following holds:

$$\forall t : (out(t) = p) \Rightarrow (trigger(t) \cap Z \neq \emptyset), \qquad (6.3)$$

where Z is the set of internal events. It means that for every transition, for which p is the input state, exists an internal event which would be necessary for firing the transition but does not occur (no active state generates it, and no transition which would generate it can fire). Let Boolean variable $\mathbf{p}$ denote activity ($\mathbf{p} = 1$) or inactivity ($\mathbf{p} = 0$) of state p in a deadlock; let Boolean variable ϵ denote availability of an internal event e. For a state p for which condition (6.3) holds:

$$(\mathbf{p} \rightarrow \bigwedge_{out(t_i)=p} \bigvee_{e_j \in trigger(t_i)} \overline{\epsilon_j}) = 1. \qquad (6.4)$$

It means, that for every state p, active in a deadlock, and for every outgoing transition ($\bigwedge_{out(t_i)=p}$), in the set $trigger(t_i)$ there is at least one event ($\bigvee_{e_j \in trigger(t_i)}$), which is not available ($\overline{\epsilon_j} = 1$).

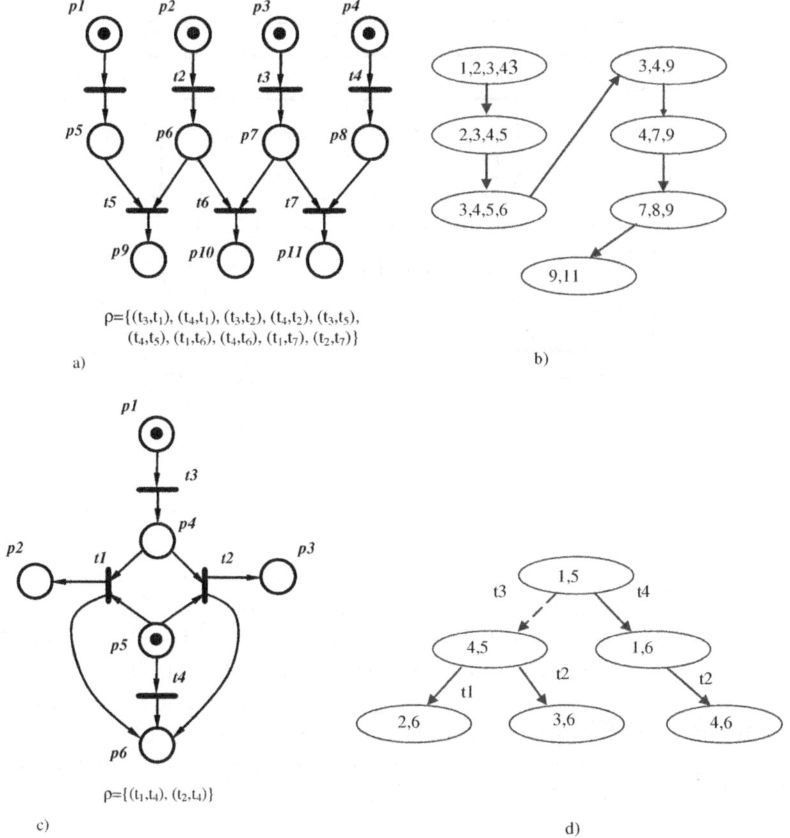

Fig. 6.6. Examples of deadlock detection of the nets with priorities

In a global deadlock, only the static and external events can be available, so
we can write that

$$\bar{\epsilon} = \bigwedge_{e \in saction(p_k)} \overline{\mathbf{p_k}}. \tag{6.5}$$

It follows from (6.4, 6.5), that a local state p satisfying (6.3) can be active in
a deadlock, if

$$(\mathbf{p} \rightarrow \bigwedge_{out(t_i)=p} \bigvee_{e_j \in trigger(t_i)} \bigwedge_{e_j \in saction(p_k)} \overline{\mathbf{p_k}}) = 1, \tag{6.6}$$

or activity of a local state in a deadlock may implicate inactivity of some
other states in this deadlock.

Constructing equation (6.6) for every local state satisfying (6.3), adding the
characteristic function of global states (which has to be generated including
events [149, 150, 151]) and specifying, by assigning 0 to the corresponding variables, that no dynamic internal events are available and no states not satisfying

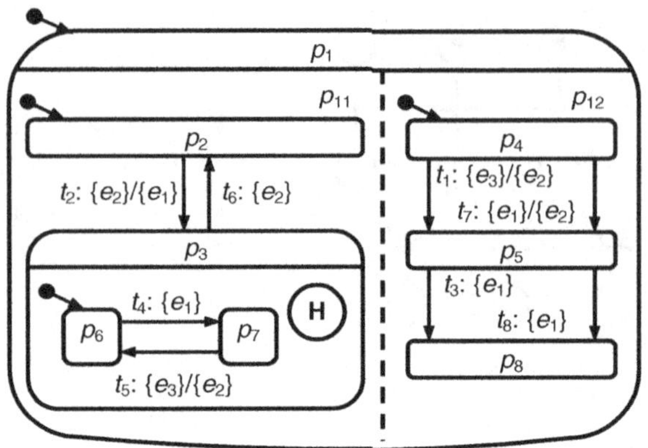

Fig. 6.7. An example of Statechart

(6.3) are active, we obtain a system of equations, with roots corresponding to all possible deadlocks.

If a system has so many global states that generation of characteristic function turns to be too time-consuming, it can be substituted by the function describing hierarchical structure of the system (*structure function*, which can be easily generated from the so-called AND-OR tree [151]). However, there is no guarantee, that the deadlocks detected in such way are reachable. Their reachability can be additionally checked by optimal simulation of the system [113].

Let us demonstrate this problem on a simple example shown in Fig. 6.7. Here $Z = \{e_1, e_2\}$. Equation 6.7 describes the structure of this Statechart:

$$(\mathbf{p_2}\overline{\mathbf{p_3}}\overline{\mathbf{p_6}}\overline{\mathbf{p_7}} \vee \overline{\mathbf{p_2}}\mathbf{p_3}(\mathbf{p_6}\overline{\mathbf{p_7}} \vee \overline{\mathbf{p_6}}\mathbf{p_7}))(\mathbf{p_4}\overline{\mathbf{p_5}}\overline{\mathbf{p_8}} \vee \overline{\mathbf{p_4}}\mathbf{p_5}\overline{\mathbf{p_8}} \vee \overline{\mathbf{p_4}}\overline{\mathbf{p_5}}\mathbf{p_8}) = 1. \qquad (6.7)$$

As far as there are no static events in this Statechart, it is enough to assign 0 to all variables corresponding to states, which do not satisfy (6.3). These states are p_4 and p_7; substituting $\mathbf{p_4} = \mathbf{p_7} = 0$ to (6.7), we obtain:

$$(\mathbf{p_2}\overline{\mathbf{p_3}}\overline{\mathbf{p_6}} \vee \mathbf{p_3}\mathbf{p_6}\overline{\mathbf{p_2}})(\mathbf{p_5}\overline{\mathbf{p_8}} \vee \overline{\mathbf{p_5}}\mathbf{p_8}) = 1. \qquad (6.8)$$

This equation has 4 solutions, corresponding to the deadlocks: $\{p_2, p_5\}$; $\{p_2, p_8\}$; $\{p_3, p_6, p_8\}$; $\{p_3, p_5, p_6\}$. The static analysis, not using the characteristic function, cannot answer the question, whether these deadlocks are reachable from the initial state. It can be checked by the dynamic analysis, as described below.

Dynamic Analysis

Let us start from modelling Statecharts by Petri nets. There are several methods of such modelling [113, 151, 154], but not all of them are adequate for our purposes. We propose the following algorithm generating a Petri net Σ for given Statechart:

Algorithm 6.6

(Here, to avoid ambiguity, Statechart transitions are denoted as t_z, and Petri net transitions as t_p.)

1. Create a place for every state p_i such that $hrc(p_i) = \emptyset$ or $history(p_i) = true$ and for every internal event e_j such that $\exists t_z : e_j \in trigger(t_z)$.
2. For every Statechart transition t_z create a set of Petri net transitions T_{tz}: for every possible combination of the active substates of $out(t_z)$, except of substates of the states with history being the substates of $out(t_z)$, a transition in T_{tz} is created; places corresponding to the states from this combination belong to the set of input places of the transition. If $hrc(out(t_z)) = \emptyset$ or $history(out(t_z)) = true$, then $|T_{tz}| = 1$. Also, $\forall t_p \in T_{tz}$:
   a) All places, corresponding to the states which will become active, when $in(t_z)$ becomes active, except of the substates of the states with history being the substates of $in(t_z)$, belong to $t_p^\bullet$.
   b) All places corresponding to the events from $taction(t_z)$ belong to $t_p^\bullet$.
   c) All places corresponding to the events from $trigger(t_z)$ belong to $^\bullet t_p$.
3. For a transition t_z between the states being the substates of a state p such that $history(p) = true$, all the Petri net transitions corresponding to t_z should have the place corresponding to p as both input and output place.
4. For every place e corresponding to a dynamic event, create a transition t_p such that $^\bullet t_p = \{e\}$, $t_p^\bullet = \emptyset$.
5. Consider every situation such that for a transition $t_z \ \exists p, e : e \in trigger(t_z)$, $e \in saction(p)$. Remove corresponding Petri net transition t_p' and for every state p such that $e \in saction(p)$ add transition t_p'' such that $^\bullet t_p'' = {}^\bullet t_p' \cap \{p\}$, $t_p''^\bullet = t_p'^\bullet \cap \{p\}$.
6. In the initial marking, the places are marked corresponding to the initially active states of the Statechart, and also the places corresponding to the states p such that $history(parent(p)) = true$, $default(parent(p)) = p$.

It is easy to see that a Petri net created as described in this section can simulate all possible evolutions of an asynchronously interpreted Statechart. The transitions added in step 4 serve for modelling the property of the events to be available only "for an instant of time" [223].

Fig. 6.8 presents the Petri net corresponding to the Statechart shown in Fig. 6.7.

Analysis of the reachability graph of the net presented in Fig. 6.7 shows that there are 3 reachable deadlocks: $\{p_2, p_5, p_6\}$; $\{p_3, p_6, p_8\}$; $\{p_3, p_5, p_6\}$, which correspond to the deadlocks of the Statechart: $\{p_2, p_5\}$; $\{p_3, p_6, p_8\}$; $\{p_3, p_5, p_6\}$. So, not all the deadlocks detected by static analysis are reachable in the system.

But simulation of the Statechart considered here leads to the conclusion, that one more deadlock - $\{p_3, p_5, p_6\}$ - is not reachable. It is directly reachable in our Petri net model from the markings $\{p_3, e_1, p_5, p_6\}$ and $\{p_3, e_2, p_5, p_6\}$, but in the first case event e_1 will be consumed by t_3 or t_4, in the second case e_2 will be consumed by t_2. The events in such situations cannot just disappear, and the

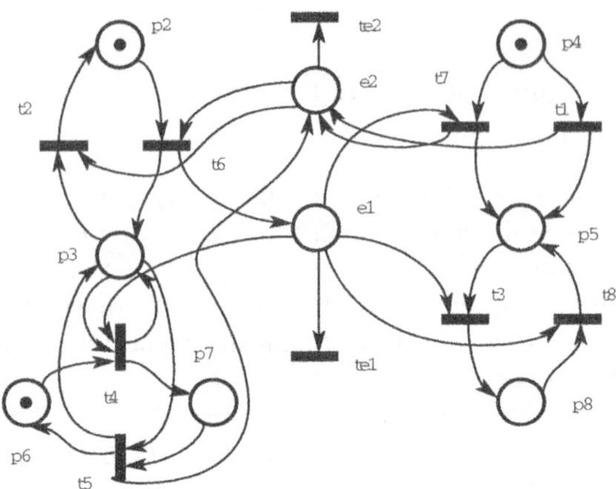

Fig. 6.8. Petri net modelling the Statechart from Fig. 6.7

state $\{p_3, p_5, p_6\}$ turns to be unreachable. It would not be the case, if t_2, t_3 and t_4 would require some external events to fire; but they do not.

What went wrong in our analysis? The Petri net model would be adequate here, if we add priorities: a transition t added in step 4 with respect of event e should have lower priority than any other transition t' having an input place common with it, if for the Statechart transition t'_z corresponding to t': $trigger(t'_z) \supseteq Z$. Intuitively, it means that an available event cannot just "disappear", if an enabled transition can consume it.

Simulation of the net of Fig. 6.8 with priorities as described above shows, that there are two reachable deadlocks - $\{p_2, p_5, p_6\}$ (corresponding to deadlock $\{p_2, p_5\}$ of the Statechart) and $\{p_3, p_6, p_8\}$.

In this example the reachability graph cannot be reduced by stubborn set method; in general case the variant of this method for the Petri nets with priorities, described in Subsection 6.1.1, can be applied here. As far as we have an adequate Petri net model and a method allowing finding of the deadlocks in Petri nets, it is possible to formulate a method of analysis in terms of Statecharts. It is convenient, however, to construct here the stubborn-like sets P_S of states and events, not of the transitions. The algorithm of constructing a set P_S for a Statechart at global state M, based on Definition 6.5, is presented below.

Algorithm 6.7

1. For a state $p \in P$ such that $M(p) = 1$ and $\exists t \in T$ such that $out(t) = p$ and t is enabled: $P_S := \{p\}$.
2. For every p, t such that $p \in P_S, out(t) = p, t$ is enabled: $P_S := P_S \cup trigger(t)$;
3. If $\exists p, t : p \in P_S, out(t) = p$, t is disabled because of lacking internal events, and there is no e in P_S such that $e \in (Z \cap trigger(t))$ and $M(e) = 0$, add such e to P_S.
4. For every e such that $e \in P_S$ and $M(e) = 1$, for every $t' \in T$ such that p' is an immediate substate of a superstate of type OR (not necessarily immediate)

of $out(t)$ and $M(p') = 1$, if such p' does not belong to P_S yet (it exists by construction).

5. For every e such that $e \in P_S$ and $M(e) = 0$, if there is no $p \in P_S$ and t such that p is an immediate substate of a superstate of type OR (not necessarily immediate) of $out(t)$, $e \in action(t)$) and $M(p) = 1$, add such p to P_S.
6. If $p \in P_S$, $M(p) = 1$ and there is enabled transition t such that $out(t) = p$, add to P_S all the active sub and superstates of type OR (not only immediate), being the immediate substates of the states of type OR, if those states do not belong to P_S yet.
7. If there is a transition t such that $trigger(t) \subseteq Z$ and there is enabled t' such that $out(t') \in P_S$ and firing of t' activates , $out(t)$: $P_S := P_S \cup trigger(t)$;
8. If there is $e \in P_S$, $M(e) = 1$, $t \in T$, $trigger(t) \subseteq Z$, $e \in trigger(t)$: add to P_S the state p such that p is an immediate substate of a superstate of type OR (not necessarily immediate) of $out(t)$ and $M(p) = 1$, if p does not belong to P_S yet.
9. Repeat 2-8 while P_S grows.

It is easy to see that by the construction P_S contains only the active states and the internal events. Steps 7, 8 are added to take into account the fact, that an event can disappear only if there is no transition able to consume it.

Deadlock detection can be performed as follows. Start from the global state M_0 (all events are initially unavailable). Then, for every considered global state M, construct P_S by Algorithm 6.7 and simulate firing of all the enabled transitions originating from the states belonging to P_S and "disappearing" of all the events belonging to P_S, which can disappear without being consumed by some active transitions. Such graph will contain all reachable deadlocks of the system.

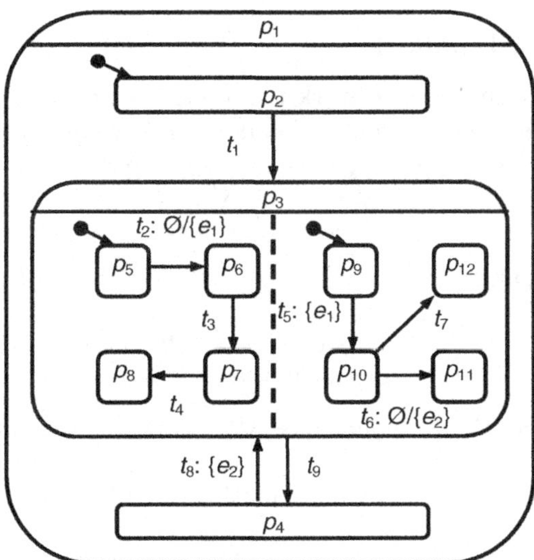

Fig. 6.9. Another example of Statechart. External events are not shown.

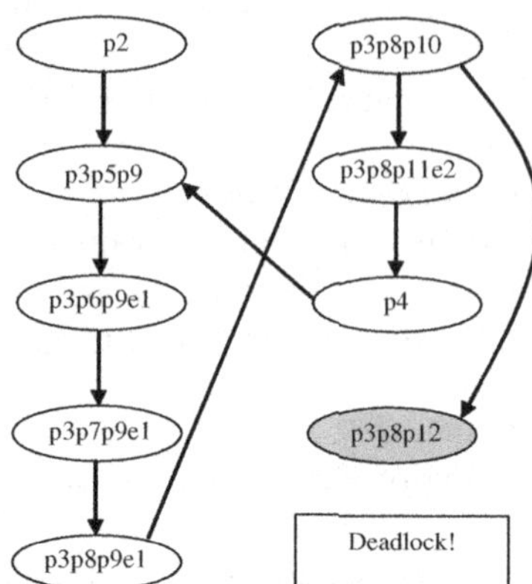

Fig. 6.10. Reduced reachability graph for the Statechart shown in Fig. 6.9

In Fig. 6.10 the reduced reachability graph for the Statechart shown in Fig. 6.9, obtained by means of the described method, is present. Full reachability graph in this case contains 15 nodes, the reduced graph contains only 9.

6.1.3 FSM Networks

FSM networks are much simpler for analysis and implementation, than the general-case Statecharts: in all global states the number of active local states is the same, and every automaton has exactly one active state. Below we discuss two tasks of verification of such networks: detection of local deadlocks and potentially unreachable local states.

Detection of Local Deadlocks

An internal event in an FSM network is absent because of a deadlock, if every FSM which can generate this event is deadlocked. For static events we can write:

$$(\bar{\epsilon} \rightarrow \bigwedge_{e \in saction(p_i^j)} \bigvee_{p_l^j \in P^j, l \neq i} \mathbf{p}_l^j) = 1, \tag{6.9}$$

and for dynamic events:

$$(\bar{\epsilon} \rightarrow \bigwedge_{e \in taction(t_i^j)} \bigvee_{p_l^j \in P^j} \mathbf{p}_l^j) = 1, \tag{6.10}$$

where $\mathbf{p}$ and ϵ are the Boolean variables with the same meaning as in equations (6.4 - 6.6).

From (6.4, 6.9, 6.10) the following equations can be constructed: for every p satisfying (6.3)

$$(\mathbf{p} \rightarrow \bigwedge_{out(t)=p} \bigvee_{e \in trigger(t)} \bigwedge_{e \in saction(p_i^j)} \bigvee_{p_l^j \in P^j, l \neq i} \mathbf{p}_k^j) = 1, \tag{6.11}$$

$$(\mathbf{p} \rightarrow \bigwedge_{out(t)=p} \bigvee_{e \in trigger(t)} \bigwedge_{e \in taction(t_i^j)} \bigvee_{p_l^j \in P^j} \mathbf{p}_l^j) = 1. \qquad (6.12)$$

It follows that activity of a local state in a deadlock implies activity of some other local states. If all the automata in the network are the Moore automata, it is sufficient to use equations (6.11); if all of them are the Mealy automata, it is sufficient to use (6.12). Equations for all the local states satisfying (6.3) together with the characteristic function specify a system of equations, having roots corresponding to the global (if all the automata participate in it) or local deadlocks.

If the characteristic function is not available, it can be replaced by the structure function, which can be built in evident way for an FSM network. However, in such case only dynamic analysis can definitively answer the question, whether a deadlock is reachable from the initial state.

In case when the deadlock is local, it is not always necessary to analyze dynamically the whole network. It may be sufficient to analyze a subsystem, selected by the Algorithm 6.8. Its input data are an FSM network and a local deadlock M_d detected by static analysis; its output is a subsystem (a set of automata) N'.

Algorithm 6.8

1. Add to N' (initially empty) all the automata, participating in the deadlock.
2. Add to N' all the automata generating events consumed by the transitions in N'.
3. Repeat step 2 while changes of N' occur.

Now for each such subsystem (of course there may exist the same subsystems for several deadlocks or the subsystems including each other; in that case there is no sense to analyze separately the subsystems included by other subsystems) deadlocks can be detected by constructing a reduced reachability graph using Algorithm 6.7.

Detection of Unreachable Local States

A more general case of a behavioral problem than a deadlock is unreachability of local states. If some local states are inactive in all the reachable global states, it can be detected from the characteristic function. This function does not, however, answer the question, whether a local state, initially reachable, can become unreachable. Check of this possibility requites another analysis method. The problem of local states reachability, in spite of its undoubted practical importance, has not been studied for FSM networks in the aspect presented here, as far as we know.

The task is formulated as follows: an FSM network is given. If there exist such global states of the network, that some states of the automata in the network are unreachable from these global states, detect such cases.

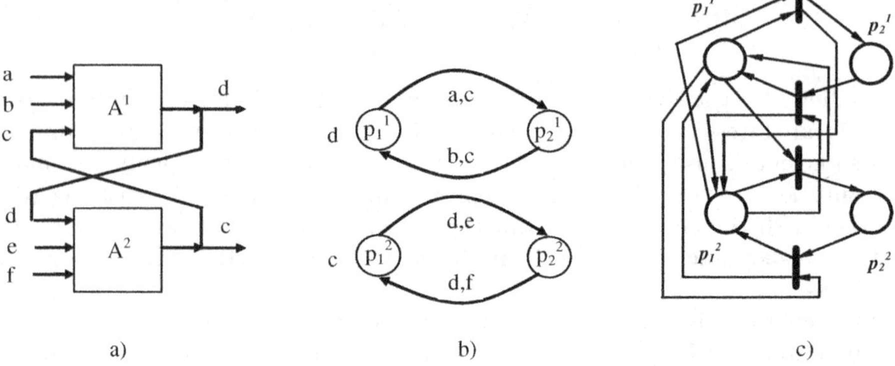

Fig. 6.11. A simple example of FSM network: a) structure of the network; b) state transition graphs of the FSMs in the network; c) modelling Petri net obtained by Algorithm 6.6.

An FSM network can be modelled by Petri net with application of Algorithm 6.6 (if only the communicating automata without hierarchy are considered, the algorithm can be simplified, as described in [116]). An example of such modelling is shown in Fig. 6.11.

Below we detect unreachable states by analyzing the modelling Petri nets. In some cases such analysis can be simplified by reduction. If there is a strongly connected subgraph of a state transition graph of an FSM such that no transitions require events from other FSMs and no states generate input events for other FSMs, the corresponding Petri net places can be replaced by a single place. In some cases a connected (not strongly) subgraph of state transition graph can also be replaced by single place, but detailed discussion of this topic remains beyond the scope of this book.

If the modelling net is live, no unreachable local states can exist in the system; but, first, the opposite is not true in general case, because steps 2 and 5 of Algorithm 6.6 can bring redundancy into the net; second, there seems to be no simple way of deciding liveness of the nets of this kind. So we propose the method which does not require liveness analysis.

The method is based on the following affirmation.

Lemma 6.9. *Let N be an FSM network and Σ - the Petri net modelling it (built using Algorithm 6.6). If there is siphon D in Σ such that there is no FSM in N for which D contains all places corresponding to it, then there exists a global state of N such that no local states corresponding to the places of D are reachable.*

Proof. If siphon D does not contain the places corresponding to all states of any FSM in the network, then there exists a global state M' of N such that no place in D is marked in the corresponding marking M of Σ. No marking which marks any place of D is reachable from M, hence no such global state is reachable in N from M', that a local state corresponding to a place in D is active.

Consider the example shown in Fig. 6.11. There is siphon $D = \{p_1^1, p_1^2\}$, and for the state in which these places are not marked, the net is deadlocked. It means that there exists such global state that p_1^1 and p_1^2 will never be reached.

Lemma 6.9 does not work in another direction, and it is possible that there are unreachable local states, but there are no siphons of the kind described in Lemma 6.9. An example is shown in Fig. 6.12 (state p_1^3 may become unreachable).

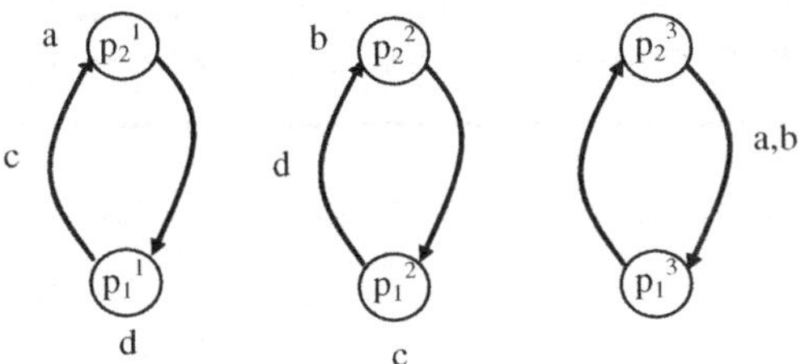

Fig. 6.12. An example of unreachability which cannot be detected by the described method

The class of the situations of unreachability, which can be detected by the method, includes all global and local deadlocks. It is shown by the next lemma.

Lemma 6.10. *Let N be an FSM network and Σ - the Petri net model of this network. If there is a set of FSMs $N' \subseteq N$ such that there is a global state M_d in which the active states of FSMs in N' constitute a deadlock, then the set D of all places corresponding to the passive states of FSMs in N' and all places corresponding to the dynamic events generated only by transitions of FSMs in N', is a siphon.*

Proof. Let $t \in {}^\bullet D$. By construction (Algorithm 6.6) it corresponds to a transition t_z of an FSM $A \in N$. Let $A \in N'$. If $M_d(out(t_z)) = 1$, then, according to the definition of deadlock, t_z cannot fire, because a static or dynamic event is not available, which can be generated only by a state or transition of an FSM $A' \in N'$. If $M_d(out(t_z)) = 0$, then $out(t_z) \in D$. In both cases there is place $p \in {}^\bullet t$ such that $p \in D$, so $t \in D^\bullet$. Let $A \notin N'$. Then, by construction, there is place p such that $p \in {}^\bullet t$, $p \in t^\bullet$. Hence ${}^\bullet D \subseteq D^\bullet$, and D is a siphon.

In Fig. 6.13 a more general example of detected situations is shown.

Siphons of a Petri net can be calculated by solving a logical equation using the Thelen's method [32, 207, 228], as described in Chapter 5.

Remark. The implicants in the Thelen's method are calculated by using a search tree; as far as we are not interested in the siphons including all the places corresponding to an FSM, the branches leading to such decisions should be pruned.

Below, the algorithm detecting possible unreachablility situations for given FSM network N is described.

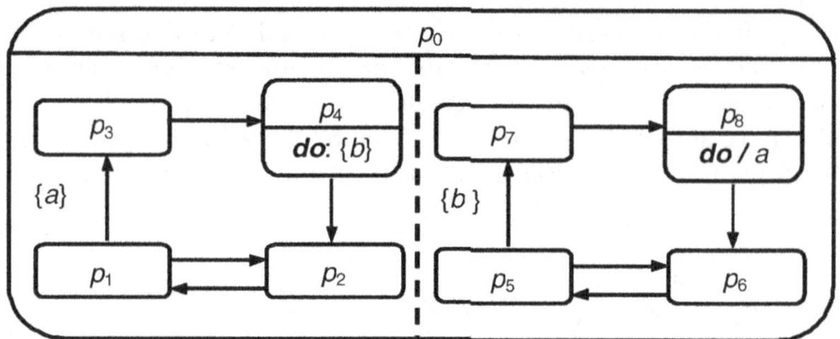

Fig. 6.13. An example of unreachability detectable by the method and not being a deadlock

Algorithm 6.11

1. Create for N a Petri net Σ by means of Algorithm 6.6.
2. If possible, perform reduction of Σ, as described above.
3. Detect the siphons of Σ by means of Algorithm 5.7, taking into account the Remark given above.
4. Every obtained siphon D (satisfying the Remark) specifies a set of global states (all the global states in which no local states corresponding to places in D are active) and a set of local states (corresponding to the places in D) which are not reachable from these global states.

Example: consider the network shown in Fig. 6.13. The corresponding Petri net (after reduction) is shown in Fig. 6.14. Place p_5^1 corresponds here to the states p_1^1, p_2^1; p_5^2 corresponds to the states p_1^2, p_2^2.

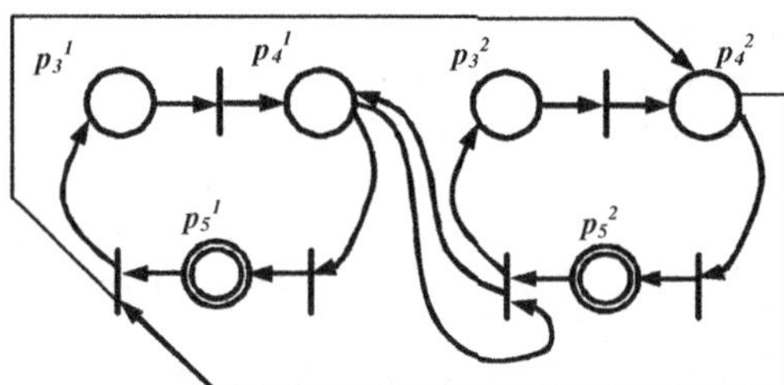

Fig. 6.14. Reduced Petri net modelling the FSM network shown in Fig. 6.13

We have the equation:

$$(\mathbf{p}_3^1 \vee \overline{\mathbf{p}_5^1} \vee \overline{\mathbf{p}_4^2})(\overline{\mathbf{p}_4^2} \vee \overline{\mathbf{p}_5^1} \vee \mathbf{p}_4^2)(\overline{\mathbf{p}_3^1} \vee \mathbf{p}_4^1)(\overline{\mathbf{p}_4^1} \vee \mathbf{p}_5^1)$$
$$\wedge (\mathbf{p}_3^2 \vee \overline{\mathbf{p}_5^2} \vee \overline{\mathbf{p}_4^1})(\overline{\mathbf{p}_4^1} \vee \overline{\mathbf{p}_5^2} \vee \mathbf{p}_4^1)(\overline{\mathbf{p}_3^2} \vee \mathbf{p}_4^2)(\overline{\mathbf{p}_4^2} \vee \mathbf{p}_5^2). \qquad (6.13)$$

After simplification we obtain:

$$(\mathbf{p}_3^1 \vee \overline{\mathbf{p}_5^1} \vee \overline{\mathbf{p}_4^2})(\overline{\mathbf{p}_3^1} \vee \mathbf{p}_4^1)(\overline{\mathbf{p}_4^1} \vee \mathbf{p}_5^1)(\mathbf{p}_3^2 \vee \overline{\mathbf{p}_5^2} \vee \overline{\mathbf{p}_4^1})(\overline{\mathbf{p}_3^2} \vee \mathbf{p}_4^2)(\overline{\mathbf{p}_4^2} \vee \mathbf{p}_5^2). \qquad (6.14)$$

The only root satisfying the Remark is: $\mathbf{p}_5^1 = \mathbf{p}_5^2 = 1$, $\mathbf{p}_3^1 = \mathbf{p}_4^1 = \mathbf{p}_3^2 = \mathbf{p}_4^2 = 0$. It means that if the network is in any global state such that the local states p_3^1, p_4^1, p_3^2, p_4^2 are not active, then these states are unreachable.

6.2 Verification of Parallel Automata Implementation

Apart from possible mistakes in high-level system specification, implementation process may also be a source of troubles. Even if the design algorithms are correct, their software implementation may turn to be not perfect (they say, there is no software without bugs). There exists the following verification problem: testing behavioral equivalence between a higher-level specification and the lower-level specification, being the result of applying design procedures to the former. There are generally two approaches to solve the problem: testing (comparing behavior of two models, representing the same system at different levels) and symbolic approach (analytical check of equivalence of these models). Of course the problem arises at various steps of implementation.

There are the following main steps of obtaining structural description of a controller from its high-level specification [40]:

- obtaining an automaton with abstract states;
- state encoding;
- obtaining a Boolean or sequent automaton;
- low-level optimization.

At all the steps some mistakes may appear. Below, we consider verification of transformation from a parallel automaton to a sequent automaton. Hierarchical structures are not considered here (each hierarchy level can be verified separately)

6.2.1 Testing Approach

Complete comparison of systems' behavior could be obtained by means of constructing the full state spaces, but then the state explosion problem arises. It can be avoided here, if the following question is answered: how to generate a reasonably short test allowing to detect incorrectness in behavior, if it exists, with high probability? Testing approach, unlike analytical one, is universal in the sense that it allows testing in similar way models of different levels, and also the hardware devices [120].

Transition coverage

In many cases, transition coverage [84] (or even coverage of selected states and transitions [129]) is considered to be sufficient. Then, we can ask how to cover all the transitions of a parallel automaton by possibly short firing sequences? In terms of Petri nets, the problem can be formulated as follows: which step sequence, allowed in the initial state, covers all transitions and is minimal by length?

For the LSα-nets, corresponding structures of the parallel automata, obtaining a coverage is simplified by the fact, that no choice performed during simulation can deadlock the net or make impossible future firing of any transition or attaining the initial marking.

Exact minimization of the sequence would require thorough investigation of the net structure and probably dealing with the full reachability graph in an explicit way. We propose a heuristic approach and make a claim that it allows to obtain "rather short" step sequences.

Heuristic (*Greedy Covering*). *At each state considered select one transition from each enabled cluster, such that:*

1. *if there are transitions in the cluster not covered yet, select one of them;*
2. *else, select transition t such that there is the shortest path in the Petri net graph to an uncovered transition starting from t, in comparison with all other transitions belonging to [t].*

Theorem 6.12. *Step sequence, obtained by applying the heuristics to a connected LSEFC-net Σ, covers all its transitions after the finite number of steps.*

Proof. Suppose the opposite. Then, as far as Σ is safe, there is a step sequence σ, constructed according to the heuristic, such that $M\sigma M$ and the set of uncovered transitions remains the same (non-empty) during the execution of σ from M. Let p belong to a cluster enabled in M; let t be the uncovered transition such that there is the shortest path in the Petri net graph from p to t (from Lemma 3.27, Σ is strongly connected, and t always exists). Denote the token in p at M as a. Suppose, that if a transition firing removes token a from a place, it appears in the output place of the transition, such that the path from this place to t in the net graph is the shortest one.

If a transition belonging to $[p]$ fires during σ, then, according to the heuristic, this is the transition from which the path to t is the shortest possible one (as far as Σ is EFC, all transitions in a cluster can be enabled only simultaneously). According to what we supposed, a appears after its firing in the place p_1 such that path from p_1 to t is shorter than from p to t. Generally, if a changes its place, the path from the place p_a, where a is situated, to t becomes each time shorter (and no uncovered transition is covered by σ, so t remains the closest uncovered transition to p_a). As far as p belongs to an enabled cluster at M, and according to the heuristic (at each step a transition in every enabled cluster fires), a does change its place already at the first step Δ_1 of σ. Let p_1 be the place containing a after Δ_1. Σ is safe, so $M(p_1) = 0$; then there is a step in σ removing token (a) from p_1. Then a appears in p_2, from which the path to t is closer, than from p and p_1. The same reasoning leads to the conclusion, that a will be removed also from p_2 during the execution of σ and will appear closer to t. As far as Σ is

Table 6.1. Simulation of interpreted PN from Fig. 6.15

step	current state	output	input	transitions firing
1	p1		start	t1
2	p2	rt	x1	t2
3	p3,p6,p11	y11,y21,y31	x11,x21,x31	t3,t6,t11
4	p4,p7,p12	y12,y22,y32	x12,x22,x32	t4,t7,t12
5	p5,p8,p13	y23,y33	x23,x33	t8,t13
6	p5,p9,p14	y24,y34	x24,x34	t9,t14
7	p5,p10,p15			t16
8	p16			t17
9	p1		start	t1
10	p2	rt	x1	t2
11	p3,p6,p11	y11,y21,y31	o1,o2,o3	t5,t10,t15
12	p5,p10,p15			t16
13	p16		r	t18
14	p2	rt		

finite, after finite number of steps (within σ) a appears in a place $p_n \in {}^\bullet t$. Then, according to the same reasoning, t (or another uncovered transition belonging to $[t]$) belongs to σ. We have come to a contradiction, because we supposed, that no uncovered transition is covered during firing of σ from M. The proof is completed.

Consider the example taken from [70], representing an algorithm for control of a driller (Fig. 6.15) (a parallel automaton of Moore type).

Table 6.1 describes simulation of this automaton, according to the proposed heuristic. For more compact representation, cells in columns "input variables" and "output variables" enumerate variables having value 1.

It is easy to see, that in this example simulation process of 13 steps covers all the transitions (selecting another transition at step 8 would lead to a sequence which is one step shorter). Other kinds of simulation would be more complicated; for example, a stubborn-set-like simulation would require 17 steps, maximal concurrent simulation - 31 step, their combination (as described in Section 3.4) - only 10 steps, but all these alternative variants would require constructing the reachability subgraphs, and in the suggested approach we deal with a linear process.

Fault Models

Transition coverage allows detecting some behavioral differences between two objects (for example, whether a transition will fire when a condition is satisfied and whether it will cause the required changes of output variables), but, of course, not all differences can be found in this way. Thorough analysis requires formalizing of possible fault models.

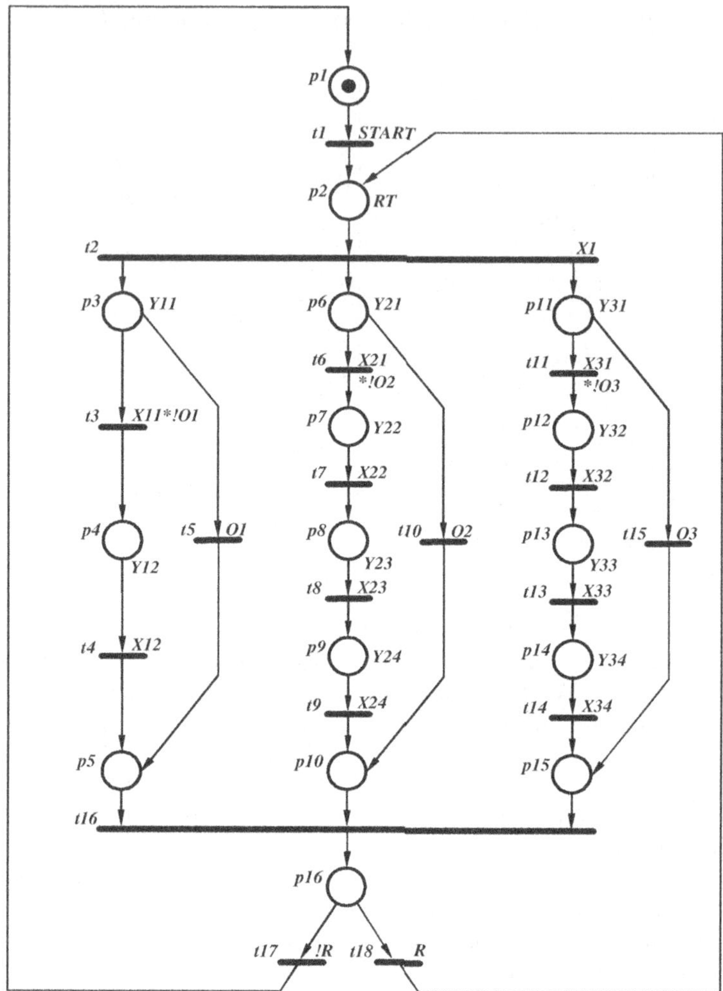

Fig. 6.15. Interpreted Petri net for a driller controller

As far as the main operation of conversion between a parallel automaton and a sequent automaton is state encoding, it is reasonable to suppose that during state encoding something may go wrong. It may lead to the next troubles during a transition firing.

1. The set of active local states required for the transition to fire is wrong.
2. A local state unnecessarily remains passive when the transition fires.
3. A local state unnecessarily remains active when the transition fires.
4. A local state unnecessarily becomes passive when the transition fires.
5. A local state unnecessarily becomes active when the transition fires.

Covering all those faults would require:

1. Checking every transition firing with all the possible combinations of active and passive local states not corresponding to its input places.
2. Checking whether - at every considered global state - a transition can fire which should not fire at that state.

The first of these 2 items requires in fact constructing full state space (there is a way to reduce it, but not radically), which we are trying to avoid. The second one, however, is much easier to satisfy; it would require no n times longer testing than with the approach considered in the previous subsection, where n is the number of transitions.

Advanced Testing

Here we propose the following addition to the testing approach considered above:

At every considered state, before testing any transition which should fire, check all the transitions which *should not* fire.

Of course it is reasonable, for minimizing testing time, to check these transitions not one-by-one, but rather joining them into the groups which should be as large as possible. For the example considered above, such testing may look as follows (Table 6.2).

For example, if there was a mistake in state encoding, and firing of transition t_4 activates place p_{16}, then it will be detected at step 16 - then the output signal RT will turn to be 1, when it should be 0. Note that the previous variant of simulation would not allow detecting this bug.

As far as not all the possible mistakes in state encoding and optimization can be detected by simulation without generation of full state space, we suppose that developing of symbolic verification methods for the transformations is also useful. This approach, based on the formal proof that the state encoding is correct, is presented below.

6.2.2 Analytical Approach

As it was mentioned above, analytical method of verification of the design process cannot be universal - a special method is needed for verification of every step of design. This subsection is dedicated to verification of the transformation from a parallel automaton to a sequent automaton. The most essential part of such transformation is the state encoding.

Below, the following task is considered: a parallel automaton A is given; a state encoding of this automaton (codes of all the local states presented by the ternary vectors of the same length) and the sequent automaton S supposed to be obtained by applying this encoding to A are also given. It is necessary to check, whether automata A and S are behaviorally equivalent. Note that here we can assume that for every transition t of A the corresponding sequent s of S is known, and that s is always a simple sequent.

Table 6.2. Advanced simulation of interpreted PN from Fig. 6.15

step	current state	output	input	transitions firing
1	p1		x1, x11, x12, x21, x22, x23, x24, x31, x32, x33, x34	
2	p1		o1, o2, o3, r	
3	p1		start	t1
4	p2	rt	start, x11, x12, x21, x22, x23, x24, x31, x32, x33, x34	
5	p2	rt	o1, o2, o3, r	
6	p2	rt	x1	t2
7	p3,p6,p11	y11,y21,y31	start, x1, x12, x22, x23, x24, x32, x33, x34	
8	p3,p6,p11	y11,y21,y31	r	
9	p3,p6,p11	y11,y21,y31	x11,x21,x31	t3,t6,t11
10	p4,p7,p12	y12,y22,y32	start, x1, x11, x21, x23, x24, x31, x33, x34	
11	p4,p7,p12	y12,y22,y32	o1,o2,o3,r	
12	p4,p7,p12	y12,y22,y32	x12, x22, x32	t4,t7,t12
13	p5,p8,p13	y23,y33	start, x1, x11, x12, x21, x22, x24, x31, x32, x34	
14	p5,p8,p13	y23,y33	o1, o2, o3, r	
15	p5,p8,p13	y23,y33	x23, x33	t8,t13
16	p5,p9,p14	y24,y34	start, x1, x11, x12, x21, x22, x23, x31, x32, x33	
17	p5,p9,p14	y24,y34	o1, o2, o3, r	
18	p5,p9,p14	y24,y34	x24, x34	t9,t14
19	p5,p10,p15		start, r, x1, x11, x12, x21, x22, x23, x24, x31, x32, x33, x34, o1, o2, o3	t16
20	p16		start, x1, x11, x12, x21, x22, x23, x24, x31, x32, x33, x34, o1, o2, o3	t17
21	p1		x1, x11, x12, x21, x22, x23, x24, x31, x32, x33, x34	
22	p1		o1, o2, o3, r	
23	p1		start	t1
24	p2	RT	start, x11, x12, x21, x22, x23, x24, x31, x32, x33, x34	
25	p2	rt	o1, o2, o3, r	
26	p2	rt	x1	t2
27	p3,p6,p11	y11,y21,y31	r	
28	p3,p6,p11	y11,y21,y31	x11, x21, x31	t3,t6,t11
29	p3,p6,p11	y11,y21,y31	o1, o2, o3	t5,t10,t15
30	p5,p10,p15		start, r, x1, x11, x12, x21, x22, x23, x24, x31, x32, x33, x34, o1, o2, o3	t16
31	p16		start, r, x1, x11, x12, x21, x22, x23, x24, x31, x32, x33, x34, o1, o2, o3	t18
32	p2	rt		

The simplest part of verification is checking whether a sequent specifies the same communication with the external world as the corresponding transition of A. For this purpose it is sufficient to check whether their input conditions and assignments to the output variables are the same. This check requires linear time[2].

Knowing the concurrency relation between the local states is necessary for verification; as far as in case of almost all state encoding methods for the parallel automata (except the one-hot encoding) [5, 39, 40, 41, 42, 186, 202, 245, 247, 249] it is also necessary, and can be calculated in polynomial time [141, 152, 186], we suppose that this information is available during verification.

For every transition $\mu - \varphi \vdash \psi \to \nu$ and corresponding sequent $\varphi f_z \vdash k_z \psi$ (the items of the list below correspond to possible troubles described in Subsection 6.2.1):

(a) there should be $f_z = \bigwedge_{p \in \mu} Qp$;
(b) there should be $(k'_z \to \bigwedge_{p \in (\nu \setminus \mu)} Qp) = 1$;
(c) there should be $k_z(\bigvee_{p \in (\mu \setminus \nu)} Qp) \equiv 0$;
(d) there should be no place $p \notin \mu$ such that $\mu \subseteq P(p)$ and $k_z Q(p) \equiv 0$;
(e) there should be no place $p' \notin \nu$ such that $(k'_z \to Q(p')) \equiv 1$.

Here k'_z is an elementary conjunction consisting of k_z and all the literals occurring in f_z which are not negated in k_z. It is easy to see that sequents $\varphi f_z \vdash k_z \psi$ and $\varphi f_z \vdash k'_z \psi$ are behaviorally equivalent, as far as the automaton is inertial and consistent (k'_z describes values of internal variables after the sequent firing).

If all mentioned conditions hold, every sequent of the sequent automaton performs changes of activity of local states exactly as it is specified by the corresponding transition of the parallel automaton.

Let us consider next example (a modified example from [249]). A parallel automaton is given (the transitions are numbered for convenience):

1. $1 - u \vdash ab \to 9$
2. $9 - \overline{u} \to 2.3$
3. $2 - \overline{v}w \vdash \overline{b}c \to 10$
4. $10 - \overline{w} \vdash b \to 11$
5. $11 - \overline{c} \to 2$
6. $2 - v \vdash \overline{a}c \to 4.5$
7. $3 - uw \vdash d \to 6$
8. $4 - \overline{u}\overline{v} \vdash a \to 12$
9. $12 - u \vdash \overline{a} \to 4$
10. $4 - u \vdash a\overline{b} \to 7$
11. $5 - \overline{v}w \vdash \overline{c} \to 8$
12. $6.7.8 \vdash \overline{a}\overline{d} \to 13$
13. $13 - \overline{w} \to 1$

---

[2] Note that the checking and the verification method described below can be applied only before any optimization procedures are applied to the sequent automaton [38, 40, 237]. Verification of optimization of the sequent automata is a task requiring special methods.

The following codes were assigned to the local states of the automaton[3]:

$$\begin{array}{c}
\begin{array}{cccccc} a & b & c & d & z_1 & z_2 \end{array} \\
\begin{array}{c}
1 \\ 2 \\ 3 \\ 4 \\ 5 \\ 6 \\ 7 \\ 8 \\ 9 \\ 10 \\ 11 \\ 12 \\ 13
\end{array}
\left[\begin{array}{cccccc}
0 & - & - & - & 0 & 0 \\
- & - & 0 & - & 1 & 0 \\
- & - & - & 0 & 1 & - \\
0 & - & - & - & 1 & 1 \\
0 & - & 1 & - & 1 & 1 \\
- & - & - & 1 & 1 & - \\
1 & 0 & - & - & 1 & 1 \\
- & - & 0 & - & 1 & 1 \\
1 & - & - & - & 0 & 0 \\
- & 0 & 1 & - & 1 & 0 \\
- & 1 & 1 & - & 1 & 0 \\
1 & 1 & - & - & 1 & - \\
- & - & - & - & 0 & 1
\end{array}\right]
\end{array} \qquad (6.15)$$

And the following sequent automaton was obtained:

$$1.\ u\overline{z_1}\bar{z_2}\bar{a} \vdash ab$$
$$2.\ \bar{u}\bar{z_1}\overline{z_2}a \vdash z_1\bar{c}d$$
$$3.\ \bar{v}wz_1\bar{z_2}\bar{c} \vdash \bar{b}c$$
$$4.\ \overline{w}z_1\overline{z_2}\bar{b}c \vdash b$$
$$5.\ z_1\overline{z_2}bc \vdash \bar{c}$$
$$6.\ vz_1\overline{z_2}\bar{c} \vdash z_2\bar{a}c$$
$$7.\ uwz_1\bar{d} \vdash d$$
$$8.\ \bar{u}\bar{v}z_1z_2\bar{a} \vdash a$$
$$9.\ uxyab \vdash \bar{a}$$
$$10.\ uz_1z_2\bar{a} \vdash a\bar{b}$$
$$11.\ \bar{v}wz_1z_2c\bar{a} \vdash \bar{c}$$
$$12.\ z_1z_2a\bar{b}\bar{c}d \vdash \overline{z_1}\bar{a}\bar{d}$$
$$13.\ \overline{w}\bar{z_1}z_2 \vdash \bar{z_2}$$

---

[3] Here the output variables a, b, c, d are used for coding, hence becoming the internal variables; z_1 and z_2 are "pure" internal variables, invisible outside the automaton.

Checking of, say, sequent 6, corresponding to transition $2 - v \vdash \overline{a}c \rightarrow 4.5$, can be performed as follows:

1. $z_1\overline{z_2}\overline{c} = Qp_2$ holds;
2. $z_2\overline{a}cz_1 = Qp_4 \wedge Qp_5$ holds;
3. $z_2\overline{a}c \wedge Qp_2 = 0$ holds;
4. there are 2 local states parallel to p_2: p_3 and p_6. For both states we have: $z_2\overline{a}c \wedge Qp_3 = z_2\overline{a}c\overline{d}z_1 \neq 0$, $z_2\overline{a}c \wedge Qp_6 = z_2\overline{a}cdz_1 \neq 0$.
5. The last condition should be checked for all local states except p_4 and p_5. For state p_1, for example, validity of the formula $z_2\overline{a}cz_1 \rightarrow \overline{a}\bar{z_1}\overline{z_2}$ should be checked. It is easy to see that the formula is not a tautology.

Checking of sequent 8, corresponding to transition $4 - \overline{u}\overline{v} \vdash a \rightarrow 12$, shows that something is wrong: $p_4 \in P(p_5)$, and $k_z \wedge Qp_5 = a \wedge \overline{a}cz_1z_2 = 0$ (condition 4 does not hold). This is caused by a mistake in state encoding: in the code of p_5 there should be "-" in position a, not 0.

An interesting addition to this method would be "decoding" of the local states: a parallel automaton and corresponding sequent automaton are given, and codes of the local states are to be restored. If the codes can be restored in noncontradictory way and the conditions described above are satisfied for all transitions, then we can be "more sure" that the implementation is correct.

As far as the symbolic method is not universal, developing the analogous methods for verification of other steps of design would be useful. Also, some encoding methods implicitly use certain knowledge about reachability of global states; applying our method for verification of such encoding may lead to wrong results. Improving the method to take into account such situations is one of the directions of our future studies.

7. Conclusion

Programming and hardware design languages, allowing to describe parallel processes, are widely used nowadays. Also, parallel formal models are widely used in the systems of computer-aided design of discrete devices. This is natural for the current level of development of system engineering and logical design, because practically every non-primitive software or hardware system, from logical control devices to multithread software applications, consists of concurrently acting objects.

However, problems of analysis and verification of parallel models are far from being solved satisfactorily. That's why developing of methods and algorithms of such verification is a task of great importance in the modern system engineering and computer science.

This book proposes several methods of solving the problems of formal verification of parallel systems. Two main classes of such systems are considered: based on Petri nets formalism and on FSM formalism (in the second case we deal with parallel compositions of FSMs). Two main approaches, based on constructing of reduced state spaces and on solving the systems of Boolean equations respectively, are used for the methods of analysis developed in this book.

Classical methods of reduced state space construction are based on selection of a subset of possible parallel evolutions of a system and simulating only the selected ones, to avoid interleaving. This approach, however, allows in general to decide only a very limited set of properties. On the other hand, in its widespread modifications it does not take into account the practically important details of parallel systems, such as interaction of the parallel processes via internal signals, which may remarkably affect the system properties. Methods of analysis of parallel systems by means of solving logical equations have not been sufficiently developed to be used in CAD systems and are not adapted to verification of the FSM-based systems.

The popular CAD systems allow analyzing the parallel models at the syntactical level, considering in fact every sequential component separately. Most of the problems caused by interaction of concurrent processes are ignored or turn to be an area of responsibility of a designer - who has no possibility to verify a complex system manually. In fact, the only way for a designer to verify certain

A. Karatkevich: Dynamic Analysis of Petri Net-Based Discrete Sys., LNCIS 356, pp. 123–125, 2007.
springerlink.com

important conditions of system correctness is simulation, which even in the best case cannot cover most of possible bugs of parallel systems.

So it seems reasonable to develop the methods of analysis of parallel systems, which can be implemented in CAD tools and would allow to perform formal verification such as deciding liveness and safeness, detecting deadlocks and unreachable states. Such methods, avoiding construction of the full state spaces, allowing however to decide the important properties, are proposed in the book.

Application of the known approaches, based on the reduced state space generation, to the subclasses of the Petri net structures, which are especially convenient for description of the parallel algorithms, allows to obtain more information on their behavior, than it is possible for the general-case Petri nets, as it is shown in Chapter 3. Further minimization of memory amount is possible; this approach can be used in cases when memory amount turns to be so high that it becomes a critical parameter (Section 3.6).

It seems to be natural to analyze the parallel systems, especially the large ones, also in parallel; therefore the methods of parallel and distributed analysis of large Petri nets are proposed (Chapter 4). The algorithm of net decomposition and analysis by a local computer network, optimized with respect to execution time, is implemented and practically verified.

A lot of information about properties of the ordinary Petri nets and the Petri net models can be obtained by solving logical equations representing the net structure. These equations can be presented and easily operated in CNF. The efficient methods of their solving and detecting by means of that the siphons and traps in Petri nets are proposed (chapter 5). On the base of these methods, the algorithm detecting deadlocks and unreachable states in FSM networks has been developed (Section 6.1.3).

Most of methods of Petri net analysis can hardly be applied to verification of the real-life designs, because, as it was mentioned above, they do not take into account some important details of these systems. So it is important to adapt the analysis methods for the various kinds of interpreted Petri nets, FSM networks, SFCs and Statecharts - that means, to the formal models sufficiently powerful to be used as detailed specifications of the discrete systems. An attempt to do it is made in Section 6.1, based on the generalized stubborn set method described in Section 3.2, the logical algebraic method presented in Chapter 5, and the modelling method from [113].

Correctness of an implementation (behavioral equivalence between high-level and low-level descriptions of the same system) also needs to be checked. Section 6.2 is dedicated to checking equivalence between a parallel automaton and a sequent automaton being a result of design operations on the parallel automaton, such as state encoding.

In the author's opinion, the main results presented in this book are the following:

- stubborn set method for the general-case parallel discrete systems;
- theoretical results on behavior of special practically important classes of Petri nets and efficient methods of analysis of Petri nets with single-token initial markings based on them;

- algorithms of memory-saving analysis of Petri nets, based on dynamic reduction of reachability graphs;
- methods of parallel and distributed analysis of large Petri nets;
- efficient method of calculation of siphons and traps in Petri nets by solving logical equations;
- methods of analysis of practically important low-level models such as interpreted Petri nets, nets with priorities and nets with inhibitor arcs;
- methods of analysis of system specifications in the form of FSM networks, sequent descriptions, SFC- and Statechart-based descriptions;
- methods of verification of transformation of parallel automata into low-level sequent descriptions;
- theoretical results on Petri nets and the stubborn set method, showing new possibilities of the method previously known;
- theoretical results on cycles in reachability graphs; a quick method of breaking of cycles in oriented graphs;
- optimization of calculation of prime implicants of Boolean functions.

Certainly, most of the research directions presented in this book require further development. The interesting and important topics of related studies, in the author's opinion, are (among others):

- developing methods of deciding well-formedness of EFC-nets and general-case Petri nets by means of constructing of reduced state spaces;
- developing the net reduction methods allowing to obtain firing sequences leading to "wrong" states;
- developing methods of dynamic analysis of continuous, hybrid and high-level Petri nets;
- design of software tools for verification of descriptions of parallel systems in VHDL, Verilog and other HDL languages, based on the dynamic and static analysis.

The approaches presented in this book, as we suppose, are not limited to the models considered here, and can be applied to analysis of other kinds of software and hardware parallel systems.

Acknowledgments

The research was supported by the KBN (Komitet Badań Naukowych, Polish State Committee for Scientific Research) grant 4T11C 006 24.

I am grateful to:

- M. Adamski and A. Zakrevskij for constant support and valuable comments on a preliminary version of this book which helped to improve it;
- A. Valmari and E. Best for consulting the author via email (to E. Best also for sending [30]);
- G. Łabiak and G. Andrzejewski for fruitful and inspiring discussions;
- J. Bieganowski for help with LaTeX;
- B. Galiński for verifying my English.

The results presented in Chapter 5 were obtained in cooperation with J. Bieganowski and A. Węgrzyn (who proposed applying Thelen's method for analysis of the Petri nets [229]). The results presented in Appendix D are based on the ideas of J. Bieganowski described in [32] (in Polish; for extended English version see [33]). Software implementation of Thelen's method with the heuristics and the computer experiments described in Appendix D were performed by J. Bieganowski. The method of distributed analysis of Petri nets described in Section 4.5 was implemented as a student project supervised by the author and consulted by T. Gratkowski, the experiments with it were performed with T. Gratkowski's essential help. The results presented in Table 4.2 were obtained as a part of a student project supervised by the author.

A. A Theorem on the Stubborn Set Method

Theorem A.1. *RRG of a bounded Petri net contains a marking, from which no deadlock is reachable, if and only if the full reachability graph contains such marking.*

Theorem A.2. *[212, 217] If no ignoring occurs in the RRG of a bounded Petri net, and the full reachability graph contains a terminal component TC, then the reduced state space contains a terminal component TC_{red} such that $TC_{red} \subseteq TC$.*

Proof of Theorem A.1.

$\Longrightarrow$ is obvious (RRG is a subgraph of full reachability graph).

$\Longleftarrow$ Let there is marking $M \in [M_0\rangle$, such that no deadlock is reachable from it. Then a terminal component TC is reachable from M in the full reachability graph. If there is no ignoring in the RRG, then, according to Theorem A.2, the RRG contains a terminal component $TC_{red} \subseteq TC$, hence containing a marking, from which no deadlock is reachable. If an ignoring occurs, then from the markings, where it occurs, no deadlock is reachable in the RRG.

A. Karatkevich: Dynamic Analysis of Petri Net-Based Discrete Sys., LNCIS 356, p. 129, 2007.
springerlink.com

B. Decyclization of the Oriented Graphs

Some methods of optimal "decyclization" of oriented graphs are described in [53, 158]. One of them is based on the fact that the adjacency matrix of an acyclic oriented graph can be transformed (by reordering columns and rows) to a strictly triangular matrix. The method reorders the matrix so that the number of non-zero elements below or on the main diagonal is minimized. Then the arcs corresponding to such non-zero elements are removed. Of course such reordering is a complex combinatorial problem.

Another method described in [53] is based on Boolean transformations: let a Boolean variable correspond to every arc of the graph, and elementary disjunction of these variables (without negation) corresponds to every cycle in the graph. Transform the obtained CNF into DNF (by multiplying the disjunctions and deleting products that subsume others) and select the shortest elementary conjunction, which will correspond to minimum feedback arc set.

By using some newer methods of Boolean transformations, this approach can be much refined. In fact, the shortest prime implicant has to be calculated here. A very efficient method of computation of the prime implicants of CNF was proposed by Thelen [207], in which the prime implicants are obtained without direct multiplying, but by using search tree and the reducing rules. In [162] a modification of the Thelen's method intended for computation of the shortest prime implicant is presented. In [33, 228] the heuristics for the Thelen's method are proposed (not affecting the results but additionally reducing the search tree). On the other hand, the task can be reduced to the deeply investigated unate covering problem [49]. So, much can be done to accelerate exact solving of the task.

Computing of all cycles in a graph requires exponential time and space in worst case; computation of the shortest prime implicant is NP, the same as the covering problem. For most practical purposes a quick approximate algorithm would be useful. Such algorithm is proposed in [111].

In this algorithm the weights are assigned to the arcs of the graph according to the formula:

$$w(e) = id(init(e)) - od(init(e)) + od(ter(e)) - id(ter(e)), \qquad \text{(B.1)}$$

A. Karatkevich: Dynamic Analysis of Petri Net-Based Discrete Sys., LNCIS 356, pp. 131–133, 2007.
springerlink.com

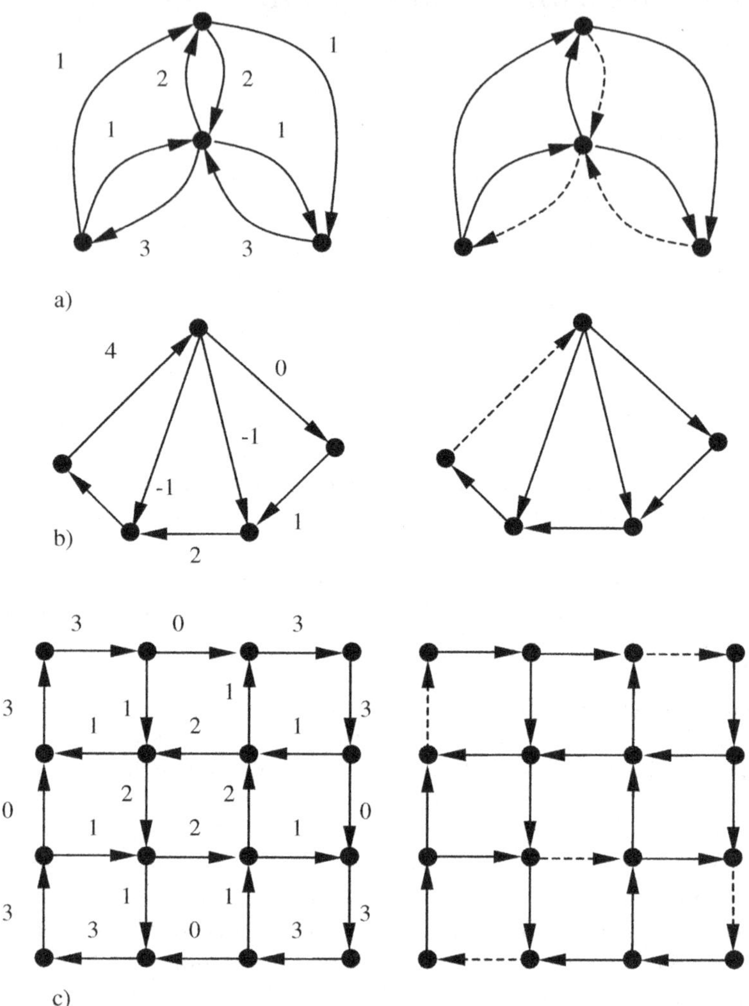

Fig. B.1. Examples of breaking cycles in oriented graphs by the described algorithm

where e is an arc, $w(e)$ is the weight, id and od are input and output degrees respectively, $init(e)$ and $ter(e)$ are the initial and terminal nodes of the arc e, correspondingly. Then the arcs are sorted (in order of nondecreasing weights) and added to the acyclic oriented graph being constructed, excluding the arcs adding of which would create a cycle. The algorithm processes each strongly connected component separately.

This process is similar to the process of building minimal spanning tree in Prim's algorithm [48], but, of course, a greedy algorithm cannot guarantee the optimal solution in this case. The intuition behind the algorithm is the following: if the initial node of an arc has many incoming arcs and few outgoing arcs, and its terminal node has many outgoing arcs and few incoming arcs, then it is likely that the arc belongs to many cycles and it is one of the few or the only one common arc of these cycles. So, it is better not to add such arc to the acyclic graph being built, that's why higher weight is assigned to it.

The time complexity of the algorithm is $((|V| + |E|)^2)$. For many examples the solutions it gives are exact or close to exact (see Fig. B.1).

C. Intersecting P-Blocks

Proof of Lemma 4.11

If $P'_{in} \cap P''_{in} \neq \emptyset$, suppose that $P'_{in} \subseteq P''_{in}$. Then if $P'_{out} \subset P''$, then Σ' is a subnet of Σ'', and if $P'_{out} \setminus P'' \neq \emptyset$, then $P'_{out} \cap P'' = \emptyset$, because Σ'' can loose tokens only through P''_{out} (by definition). It means that $P''_{out} \subset P'$, hence (from (4.1)) $P'' \subset P'$, $P'_{in} = P''_{in}$ (or $P''_{out} = P'_{in}$, but then $M''_{in} > M''_{out}$, which contradicts (4.4)), and Σ'' is a subnet of Σ'. So if $P'_{in} \cap P''_{in} \neq \emptyset$, then neither $P'_{in} \subseteq P''_{in}$, nor $P''_{in} \subseteq P'_{in}$. If $P'_{out} \subset (P'' \setminus P''_{out})$, then Σ'' can get tokens not only through P''_{in}, which contradicts (4.1). If $P'_{out} \setminus (P'' \setminus P''_{out}) \neq \emptyset$ and $P'_{out} \cap (P'' \setminus P''_{out}) \neq \emptyset$, then Σ'' can loose tokens not only through P''_{out}, which also contradicts (4.1). If $P''_{out} \cap (P' \setminus P'_{out}) \neq \emptyset$, then all the variants of intersection between P''_{out} and P' lead to a contradiction in similar way; the variants $P'_{out} \subseteq P''_{out}$ and $P''_{out} \subseteq P'_{out}$ are also contradictory. So if $P'_{in} \cap P''_{in} \neq \emptyset$, then: $(P'_{in} \setminus P''_{in}) \neq \emptyset$, $(P''_{in} \setminus P'_{in}) \neq \emptyset$, $(P'_{in} \setminus P''_{in}) \cap P'' = \emptyset$, $(P''_{in} \setminus P'_{in}) \cap P' = \emptyset$, $(P'_{out} \setminus P''_{out}) \neq \emptyset$, $(P''_{out} \setminus P'_{out}) \neq \emptyset$, $(P'_{out} \setminus P''_{out}) \cap P'' = \emptyset$, $(P''_{out} \setminus P'_{out}) \cap P' = \emptyset$. Let us call it the first variant.

If $P'_{in} \cap P'' = \emptyset$, then $P'_{out} \subseteq (P'' \setminus P''_{i/o})$; considering of other variants leads to contradiction with Definition 4.5 or to conclusion that Σ'' is a subnet of Σ'. Let us call it the second variant. There is no need to consider separately the third of possible variants $(P'_{out} \cap P'' = \emptyset, P'_{in} \subseteq (P'' \setminus P''_{i/o}))$, because it is symmetrical with the second one.

So there are only two possible variants of intersection between P-blocks, illustrated by Fig. 4.8.

The first variant (Fig. 4.8a): there is no transition t such that $({}^{\bullet}t \subset P' \setminus P'') \wedge (t^{\bullet} \subset P'' \cap P')$, $(t^{\bullet} \subset P' \setminus P'') \wedge ({}^{\bullet}t \subset P'' \cap P')$, $({}^{\bullet}t \subset P'' \setminus P') \wedge (t^{\bullet} \subset P' \cap P'')$ or $(t^{\bullet} \subset P'' \setminus P') \wedge ({}^{\bullet}t \subset P' \cap P'')$, because the opposite would contradict Definition 4.5. So, the subnets Σ_1, Σ_2 and Σ_3 satisfy (4.1); it means that these subnets are independent of each other (there is no synchronization between them). Then, if Σ' and Σ'' satisfy conditions (4.3,4.4), then Σ_1, Σ_2 and Σ_3 satisfy them, too. It is easy to see, that if (4.2) holds for Σ' and Σ'', then either all the places in $\Sigma'_{in} \cap \Sigma''_{in}$ are marked in M_0, or neither of them is marked, and Σ_1, Σ_2 and Σ_3 satisfy (4.2). So, for Σ_1, Σ_2 and Σ_3 conditions (4.1-4.4) hold, and they are the P-blocks.

A. Karatkevich: Dynamic Analysis of Petri Net-Based Discrete Sys., LNCIS 356, pp. 135–136, 2007.
springerlink.com

The second variant (Fig. 4.8b): there is no transition t such that $(^\bullet t \subset P' \setminus P'') \wedge (t^\bullet \subset (P'' \cap P') \setminus P''_{in})$, $(t^\bullet \subset P' \setminus P'') \wedge (^\bullet t \subset P'' \cap P')$, $(^\bullet t \subset P'' \setminus P') \wedge (t^\bullet \subset P' \cap P'')$ or $(t^\bullet \subset P'' \setminus P') \wedge (^\bullet t \subset (P'' \cap P') \setminus P'_{out})$, because the opposite would contradict Definition 4.5. So, the subnets Σ_1, Σ_2 and Σ_3 satisfy (4.1); communication between them exists only through P''_{in} and P'_{out} ($P_{1in} = P'_{in}$, $P_{1out} = P_{2in} = P''_{in}$, $P_{2out} = P_{3in} = P'_{out}$, $P_{3out} = P''_{out}$). Then, if Σ' and Σ'' satisfy conditions (4.3, 4.4), then Σ_1, Σ_2 and Σ_3 satisfy them, too. If (4.2) holds for Σ' and Σ'', then at M_0 either all the places in P'_{in} and no other places in $P' \cup P''$ are marked at M_0, or neither of places in $P' \cup P''$ are marked; so Σ_1, Σ_2 and Σ_3 satisfy (4.2). So, for Σ_1, Σ_2 and Σ_3 conditions (4.1-4.4) hold, and they are the P-blocks.

It is also easy to show, that in both variants Σ_4 satisfies (4.1-4.4), so Σ_4 is also a P-block. The proof is finished.

D. Improvements of Thelen's Prime Implicant Method

D.1 Introduction

A lot of tasks related to computer-aided logical design require calculation of prime implicants of a Boolean function, which is often represented as a product of sums (conjunctive normal form). The most known of such tasks is the two-level minimization of Boolean functions. Most classical and modern methods of minimization (especially exact) require calculation of all the prime implicants, from which a subset representing minimal DNF is then selected. For example, the most widely used minimization program ESPRESSO uses this approach [166].

It is worth mentioning, however, that some new efficient methods have been developed in order to avoid generation of all prime implicants [50, 63, 166]. On the other hand, new variants of minimization methods requiring all the prime implicants are still being developed [166, 189, 190]. And there are many other applications of a method of prime implicants generation. For example, calculation of complement of a Boolean function (in DNF), or transformation of a Boolean equation from CNF to DNF. And vice versa - as far as, due to de Morgan's laws, transformation from DNF to CNF can be performed by a transformation from CNF to DNF. One more application is detecting siphons and traps in a Petri net, as described in Chapter 5. Generally, solutions of a logical equation can be easily obtained from prime implicants of its left part, if the right part is 1.

There are also tasks, which can be solved by calculating shortest prime implicant or prime implicants satisfying certain conditions. In [162] several such tasks from the area of logical design are discussed. Tasks of unate and binate covering can be easily represented as logical expressions in CNF and are usually solved by one of two approaches: BDD-based [64] or branch and bound, for which shortest prime implicant would correspond to the optimal solution [49]. The same is true for some graph problems, such as decyclisation of graphs [111]. The problem of detecting deadlocks in FSM networks can be reduced to the problem of generating a subset of prime implicants (see subsection 6.1.3). The approach discussed

A. Karatkevich: Dynamic Analysis of Petri Net-Based Discrete Sys., LNCIS 356, pp. 137–144, 2007.
springerlink.com   © Springer-Verlag Berlin Heidelberg 2007

in this appendix can be applied (directly or with some modifications) to the whole range of problems mentioned above.

Several algorithms for generation of prime implicants are known. The method of Nelson [174], probably historically first such method for CNF, is based on straightforward multiplying of the disjunctions and deleting the products that subsume other products. More efficient methods are known: an algorithm based on a search tree, proposed by B. Thelen [207], and recursive method described in [166]. Comparison of these two methods is beyond the scope of this work.

Execution time of the Thelen's method depends remarkably on the order of clauses and literals in the expression. Hence we may suppose that some re-ordering of the expression will increase efficiency of the algorithm. As far as the search tree in the Thelen's method is reduced by means of certain rules (described below), it is difficult to evaluate a priori the effects of different vari-ants of reordering. It is reasonable to use a heuristic approach and to verify the heuristics statistically. Some of such heuristics are described in [161, 162].

This appendix describes some new heuristics, their analysis and comparison with other known heuristics. Experiments are performed by using the randomly generated samples; optimal combination of the heuristics is formulated on the basis of experimental results.

D.2 Thelen's Method

Thelen's prime implicant method is based on the Nelson's method [174], who has shown, that all the prime implicants of a Boolean function in conjunctive form can be obtained by its transformation into disjunctive form. Nelson's transfor-mation is very time- and memory-consuming, because all intermediate products should be kept in memory, and their number grows exponentially.

Thelen's method transforms CNF into DNF in much more efficient way. It requires linear memory for transformation and additional memory for calculated prime implicants. The subsuming products are not kept in memory. A search tree is built, such that every level of it corresponds to a clause of the CNF, and the outgoing arcs of a node correspond to the literals of the disjunction. Conjunction of all the literals corresponding to the arcs at the path from the root of the tree to a node is associated with the node. Leaf nodes of the tree are the elementary conjunctions being the prime implicants of the expression or the implicants subsuming the prime implicants calculated before. A sample tree is shown in Fig. D.1.

The tree is searched in DFS order, and several pruning rules are used to minimize it. The rules are listed below.

R_1 An arc is pruned, if its predecessor node-conjunction contains the complement of the arc-literal.

R_2 An arc is pruned, if another non-expanded arc on a higher level still exists which has the same arc-literal.

R_3 A disjunction is discarded, if it contains a literal which appears also in the predecessor node-conjunction.

The rules listed above are based on the following laws of Boolean algebra:

$$a \wedge a = a; \qquad a \vee a = a; \qquad \text{(D.1)}$$
$$a \vee a \wedge b = a; \qquad a \wedge (a \vee b) = a; \qquad \text{(D.2)}$$
$$a \wedge \overline{a} = 0; \qquad a \vee \overline{a} = 1; \qquad \text{(D.3)}$$
$$a \wedge 0 = 0; \qquad a \vee 1 = 1; \qquad \text{(D.4)}$$
$$a \vee 0 = a; \qquad a \wedge 1 = a. \qquad \text{(D.5)}$$

Rules R_1 and R_3 follow immediately from (D.2) and (D.3). Rule R_2 provides that the implicants associated with the leaf nodes, if they are not prime, subsume the implicants calculated before. It means that the first calculated implicant is always prime. An arc at level i with arc-literal x, such that there is a non-expanded arc with the same arc-literal at level j higher than i, is pruned by it. An implicant obtained by expanding this arc would be at least one literal shorter than the implicant which would be obtained without applying rule R_2. As far as the path comes twice through literal x (at the levels i and j), according to (D.1), (D.2), the longest of these two implicants subsumes the shortest one. Hence the first calculated implicant cannot subsume the implicants calculated later, but it can be subsumed by them. So, applying rule R_2 allows to check whether an implicant is simple immediately after its calculation. It is enough to compare it with all the implicants calculated before. Due to this property the algorithm is less memory-consuming, because only prime implicants are kept.

H.-J. Mathony [162] has proposed the additional fourth reduction rule, which reduces the search tree up to 25%.

R_4 An arc j is pruned, if another already expanded arc i with the same arc-literal exists on a higher level v and if rule R_2 was not applied in the subtree of arc i with respect of arc l on level v which leads to arc j.

Unfortunately, using this rule complicates the algorithm remarkably, because additional information on applying rule R_2 has to be kept. Additional reduction makes smaller the probability of appearing of the non-prime implicants at the leaf nodes. There is also no guarantee that such implicants will not appear, and still it is necessary to perform checking, the same as in case of the tree built using only 3 pruning rules. Next expression is an example, for which non-prime implicants still appear even if all 4 rules are used: $(x \vee y)y(y \vee z)z(x \vee z)$ (Fig. D.3).

D.3 Heuristics for Thelen's Method

One possible way of reducing the search tree is sorting the disjunctions by their size in ascending order.

Heuristic 1 (*Sort by Length* [162]). *Choose disjunction u_j with the smallest number of literals.*

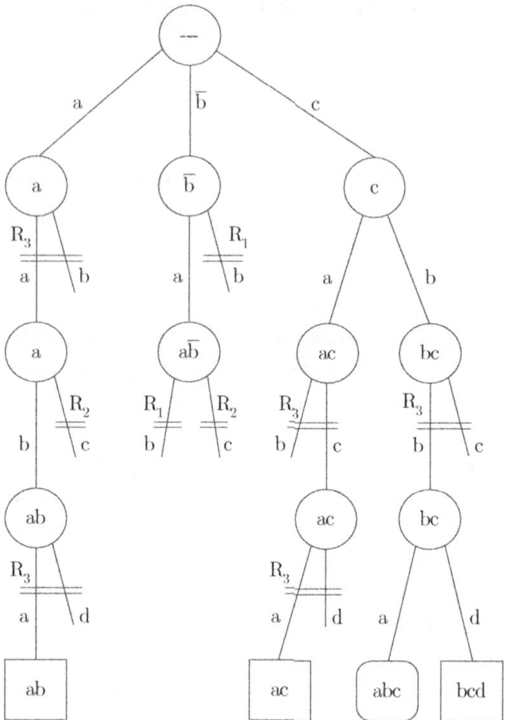

Fig. D.1. An example of the tree for Boolean formula: $(a \vee \overline{b} \vee c)(a \vee b)(b \vee c)(a \vee d)$

Effect of this heuristics can be illustrated with a complete search tree (without arc pruning). Its size (number of nodes) can be calculated according to the formula:

$$|V| = 1 + \sum_{i=1}^{n} \prod_{j=1}^{i} J_j, \tag{D.6}$$

where J_j is the number of literals in clause number j. Let a formula consist of 5 clauses, each having a different number of literals, from 2 to 6. If they are sorted from maximal to minimal length, the complete search tree will contain 1237 nodes; if sorted from minimal to maximal this tree will contain only 873 nodes. In the second case it is 30% smaller. As we see, sorting of clauses influences the tree size remarkably. Of course for the reduced search trees relation may differ.

Now let us turn to the pruning rules. Note that every rule can be implemented only if the disjunction under consideration contains the same variables as the disjunctions corresponding to the predecessor nodes. It means that if next disjunction contains the variables which appear in the previous disjunctions, there are possibilities of reduction at that level; and there is no possibility of reduction for the new variables. So we may suppose that sorting of the clauses according to the variables may also lead to tree reduction. The similar effect is used here as in the case of sorting by length: disjunctions containing many repeating variables allow to reduce the tree remarkably, and if such reduction can be performed not far form the root, the tree will be growing slower. So the following heuristics reorder the disjunctions in such a way that minimal number of new variables appear at every following level of the search tree.

Heuristic 2a (*Sort by Literals*). *Choose disjunction u_j with the minimal number of literals that do not appear in the disjunctions chosen before.*

Heuristic 2b (*Sort by Variables*). *Choose disjunction u_j with the minimal number of variables that do not appear in the disjunctions chosen before.*

The only difference between these two variants is that heuristic 2a compares clauses according to literals and heuristic 2b according to variables. This means that $\overline{x_i}$ and x_i are two different items for heuristic 2a but not for heuristic 2b. Average results of these heuristics are similar, but there are examples where the tree reduction differs a lot. So it is possible to obtain better results by selecting most effective heuristic for every example. It will be the subject of forthcoming research.

The effect of heuristics 2a and 2b is comparable to the effect of heuristic 1, but we may say, that it is more "intelligent", which is confirmed by statistical analysis. The results of computer experiments are presented in Table D.1, and as far as heuristics 2a and 2b give similar average results, only one of them is presented.

Reordering literals in clauses also affects the search tree, because it changes order of generated prime implicants and may make rule R_2 applicable or not at certain levels of the tree.

Two following heuristics reorder literals in clauses. The first of them allows quick calculation of the shortest prime implicant, and the second heuristic reduces search tree when it is necessary to calculate all prime implicants.

Heuristic 3 (*Sorting literals*[162]). *Choose literal v_i with the maximum frequency in the non-expanded part of the expression.*

Sorting the literals according to heuristic 3 leads to the situations, in which rule R_3 will be more often applicable for the arcs at the left side than at the right side. Hence, probably the first calculated prime implicants will be the shortest one. Now by using branch-and-bound method most of other arcs will be cut in several steps.

If all prime implicants have to be generated, the ordering should be different: in such case it is better to generate shortest implicants later (reverse order of literals given by heuristic 3). That reduces probability of appearing non-prime implicants at the leaf nodes. As far as due to rule R_2 an implicant can subsume only the implicants calculated before, if the implicants calculated later are in most cases shorter than those calculated earlier, then chance of subsuming is small. The following heuristic is a reversion of heuristic 3.

Heuristic 4 (*Reordering Literals*). *Choose literal v_i with the minimum frequency in the non-expanded part of the expression.*

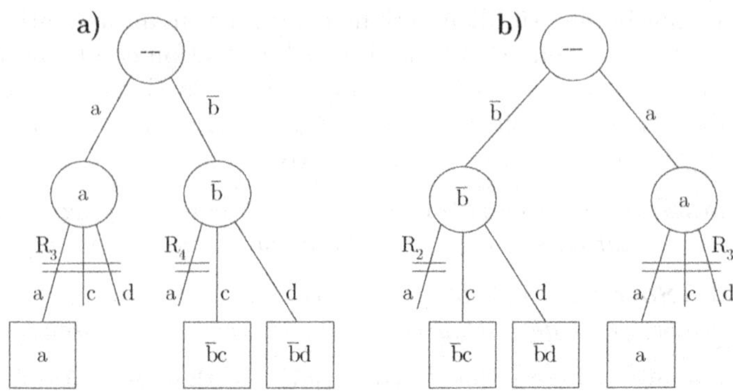

Fig. D.2. An example of the tree, in which effects of heuristic 4 and rule R_4 are the same

In many cases (but not always) effects of rule R_4 and heuristic 4 are very similar. Rule R_4 prunes an arc, if at a higher level there is a non-expanded arc with the same arc-literal (let it be a). It means that at the level i literal a is not the last literal in the clause. Let literal $\bar{b}$ be situated after a in the clause. From the arc corresponding to the literal $\bar{b}$ there is a path to the node under consideration at level l. If literal a would be the last in the clause, instead of R_4, rule R_2 would be applicable with the same effect (Fig. D.2).

We may also state that if literal a appears at level i and also at a lower level l (in clauses u_i and u_l ($i > l$)), then if b does not appear in the clauses with numbers greater than i, after applying heuristic 4 in the clause u_i literal b will appear before a and R_2 will be applicable instead of R_4. But if b appears in the clauses with numbers greater than i, this effect will not always occur.

Here is an example of heuristic 4:

$$(a \vee \bar{b})(a \vee \bar{c} \vee d).$$

After applying the heuristic:

$$(\bar{b} \vee a)(a \vee \bar{c} \vee d).$$

Such ordering of literals causes that the arc leading to non-prime implicant $\bar{b}a$, which in the first case could be pruned only by applying rule R_4, will now be pruned by rule R_2.

Another example:

$$(\bar{a} \vee b \vee c)(a \vee \bar{b})(a \vee \bar{c} \vee d)(\bar{b} \vee c).$$

In this case heuristic 4 does not change ordering of the literals. Literals a and $\bar{b}$, appearing in clause 2, appear in the next clauses with the same frequency, and without applying rule R_4 the algorithm will generate a non-prime implicant $\bar{b}a$.

On the other hand, it may happen that reordering of literals by heuristic 4 allows pruning the arcs which would not be pruned by rule R_4. It is possible

Table D.1. Results of computer experiments

Bool. form.	K	STD	H1		H2a		H4		H2a + H4		R4	
		V	V	%	V	%	V	%	V	%	V	%
20×30	144	42194	7680	18.2	7522	17.8	2536	6.0	702	1.7	7856	18.6
20×26	145	7755	9504	122.6	6391	82.4	2596	33.5	2191	28.3	4268	55.0
20×22	12	1422	645	45.4	599	42.1	247	17.4	283	19.9	247	17.4
20×22	105	2486	3166	127.4	3166	127.4	702	28.2	1067	42.9	702	28.2
20×21	70	3024	1543	51.0	1435	47.5	816	27.0	738	24.4	816	27.0
20×21	28	1346	671	49.9	635	47.2	333	24.7	247	18.4	333	24.7
20×23	36	4356	1638	37.6	1575	36.2	705	16.2	654	15.0	705	16.2
20×23	16	2650	660	24.9	826	31.2	881	33.2	391	14.8	881	33.2
20×21	117	3418	4181	122.3	4181	122.3	773	22.6	1139	33.3	773	22.6
25×31	946	51275	26030	50.8	26075	50.9	6365	12.4	2910	5.7	6374	12.4
25×29	144	55608	4222	7.6	1770	3.2	2283	4.1	438	0.8	2859	5.1
25×32	560	63234	22826	36.1	18831	29.8	4170	6.6	2125	3.4	4170	6.6
25×26	91	6838	4603	67.3	4109	60.1	1917	28.0	970	14.2	1919	28.1
average:				**58.5**		**53.7**		**20.0**		**17.1**		**22.7**

because the rules R_2 and R_4 are not completely symmetrical. Rule R_4 has additional condition which is absent in rule R_2. This condition may block applying rule R_4. But if the literals can be reordered in such a way that rule R_2 will be applicable, then such an arc will be pruned. Fig. D.3 illustrates this situation.

In the tree for expression $(x \vee y)x(y \vee z)z(x \vee z)$ (Fig. D.3a) rule R_4 cannot be applied because the condition is not satisfied (in the left subtree rule R_2 was applied), that's why in the right subtree a non-prime implicant xyz appears. Heuristic 4 changes the expression into the form: $(y \vee x)x(y \vee z)z(x \vee z)$. Now, the non-prime implicant does not appear, because the arc leading to this implicant is pruned by rule R_2 (Fig. D.3b).

Experiments demonstrate that both heuristic 4 and rule R_4 efficiently and almost at the same extent reduce the number of generated non-prime implicants. Rule R_4 is more difficult for implementation and increases necessary memory amount. It seems that applying heuristic 4 is more reasonable because allows to obtain similar effect with less effort.

Results of computer experiments are summarized in Table D.1. For the tests, the randomly generated Boolean expressions were used. In the first column number of variables and number of clauses of an expression are given (e.g. 20×18). V denotes tree size (number of nodes); K denotes number of prime implicants. The column '%' shows percentage of the tree size for every heuristic, in respect of the size in case when no heuristic is used.

The experiments show that it is better to sort disjunctions according to heuristic 2a, and literals in the disjunctions according to heuristic 4.

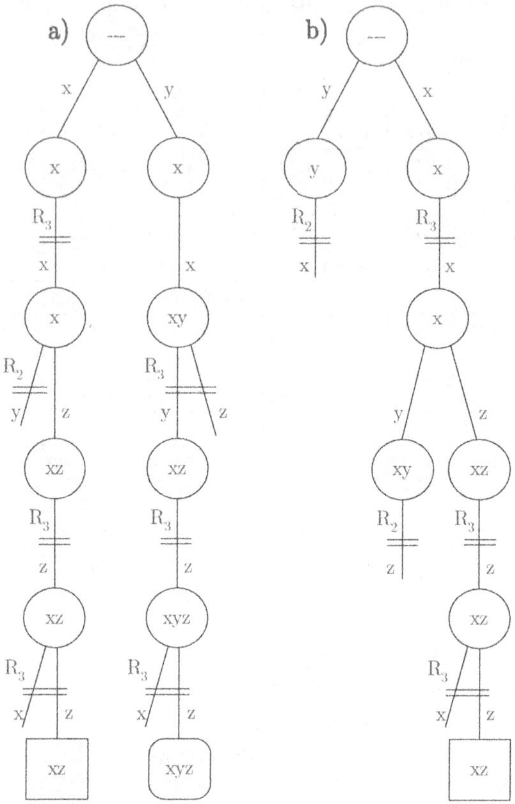

Fig. D.3. An example of the tree, in which there are differences between heuristic 4 and rule R_4

References

1. P. A. Abdulla, S. P. Iyer, and A. Nylén. Unfoldings of unbounded Petri nets. In *Proceedings of the International Workshop on Computer Aided Verification*, pages 495–507, 2000.
2. P. A. Abdulla, B. Jonsson, M. Kindahl, and D. Peled. A general approach to partial order reductions in symbolic verification (extended abstract). In *Proceedings of the International Workshop on Computer Aided Verification*, pages 379–390, 1998.
3. M. Adamski. Realizacja sieci Petri z wykorzystaniem PLA. In *Krajowa Konf. Teoria Obwodów i Układy Elektroniczne*, pages 455–459, 1981.
4. M. Adamski. *Projektowanie układów cyfrowych systematyczna metoda strukturalna*. Wyższa Szkoła Inżynierska w Zielonej Górze, 1990.
5. M. Adamski. Parallel controller implementation using standard PLD software. In *FPGAs : International Workshop on Field Programmable Logic and Applications*, pages 296–304. Abingdon EE&CS Books, 1991.
6. M. Adamski, A. Karatkevich, and M. Węgrzyn, editors. *Design of Embedded Control Systems*. Springer-Verlag, New York, 2005.
7. M. Adamski, J. L. Monteiro, W. Fengler, and A. Wendt. A distributed Petri net-based discrete controller system. In *Proceedings of the Conference on Automatic Control - Control'96*, volume 2, pages 777–782, Oporto, 1996.
8. M. Adamski and M. Węgrzyn. Field programmable implementation of concurrent state machine. In *Computer - Aided Design of Discrete Devices: Proceedings of the International Conference*, volume 1, pages 4–12, Minsk, 1999.
9. M. Adamski, M. Węgrzyn, and P. Wołanski. A VHDL based approach to logic controllers design. In *Proc. of Int. Conference Programmable Devices and Systems*, pages 9–16, 1998.
10. R. Alur, R. K. Brayton, T. A. Henzinger, S. Qadeer, and S. K. Rajamani. Partial-order reduction in symbolic state space exploration. In *Proceedings of the International Workshop on Computer Aided Verification, Lecture Notes in Computer Science 1254*, pages 340–351. Springer-Verlag, 1997.
11. G. Andrzejewski. Hierarchical Petri nets as a representation of reactive behaviors. In *Proceedings of the International Conference on Applied Computer Systems*, volume 2, pages 145–154, Szczecin, Poland, 2001.
12. G. Andrzejewski. *Programowy model interpretowanej sieci Petriego dla potrzeb projektowania mikrosystemów cyfrowych*. PhD thesis, Uniwersytet Zielonogórski, 2003.

13. G. Andrzejewski. Hierarchical Petri nets for digital controller design. In M. Adamski, A. Karatkevich, and M. Węgrzyn, editors, *Design of Embedded Control Systems*, pages 27–36. Springer-Verlag, New York, 2005.

14. G. Andrzejewski and A. Karatkevich. Interpreted Petri nets in system design. In *Proceedings of the 10th International Conference on Machine-Building and Technosphere in XXI Century*, volume 4, pages 7–10, Donetsk, Ukraine, 2003.

15. G. Andrzejewski and A. Karatkevich. Program model of hierarchical Petri net. In *Proc. of the 2nd International Workshop on Discrete - Event System Design, DESDes 04*, pages 9–14, Dychów, Polska, 2004.

16. G. Andrzejewski and A. Karatkevich. Interpreted hierarchical Petri nets in digital controller design. *Radioelektronika i Informatika*, (1):74–79, 2005.

17. M. Auguin, F. Boeri, and C. Andre. Systematic method of realization of interpreted Petri nets. *Digital Processes*, (6):55–68, 1980.

18. F. Balarin, editor. *Hardware-Software Co-Design of Embedded Systems. The POLIS Approach*. Kluwer Academic Publishers, 1997.

19. Z. Banaszak, J. Kuś, and M. Adamski. *Sieci Petriego. Modelowanie, sterowanie i synteza systemów dyskretnych*. Wyższa Szkoła Inżynierska, Zielona Gora, 1993.

20. J. Baranowski. *Metody syntezy układów cyfrowych opisanych sieciź Petri*. PhD thesis, Politechnika Ślźska, Gliwice, 1982.

21. K. Barkaoui and M. Minoux. A polynomial-time graph algorithm to decide liveness of some basic classes of bounded Petri nets. In *Proceedings of the 13th International Conference on Application and Theory of Petri Nets, Lecture Notes in Computer Science*, volume 616, pages 62–75. Springer-Verlag, 1992.

22. B. Baumgarten. *Petri-Netze: Grundlagen und Anwendungen*. Spektrum, Akad. Verlag, 1996.

23. H. Belhadj, L. Gerbaux, M. Bertrand, and G. Saucier. Proc. of the IFIP WG10.2/WG10.5 Workshops on Specification and synthesis of communicating finite state machines. In *Synthesis for Control Dominating Circuits*, pages 91–102. Elsevier, North-Holland, 1993.

24. G. Berthelot. Checking properties of nets using transformation. In *Advances in Petri Nets'85, Lecture Notes in Computer Science*, volume 222, pages 19–40. Springer-Verlag, 1986.

25. G. Berthelot and C. Roucairol. Reduction of petri nets. In *Mathematical Foundations of Computer Science, Lecture Notes in Computer Science*, volume 45, pages 202–209. Springer-Verlag, Berlin, 1976.

26. G. Berthelot, C. Roucairol, and R. Valk. Reduction of nets and parallel programs. In *Lecture Notes in Computer Science*, volume 84, pages 277–290. Springer-Verlag, 1980.

27. E. Best. Petri net semantics of priorities. In *Extended Abstracts of Concurrency and Compositability*, pages 11–14, S. Miniato, 1990.

28. E. Best, L. Cherkasova, J. Desel, and J. Esparza. Characterisation of home states in free choice systems. In *Semantics for Concurrency. Proceedings of the International BCS-FACS Workshop*, pages 16–20, London, UK, 1990. Springer-Verlag.

29. E. Best and J. Desel. Partial order behaviour and structure of Petri nets. *Formal Aspects of Computing*, 2:123–138, 1990.

30. E. Best and M. Koutny. Petri net semantics of priority systems. *Theoretical Computer Science*, (96):175–215, 1992.

31. G. Bhat and D. Peled. Adding partial orders to linear temporal logic. In *Proceedings of the 8th International Conference CONCUR'97, Lecture Notes in Computer Science*, volume 1243, pages 119–134, Berlin, Germany, July 1997. Springer-Verlag.

32. J. Bieganowski and A. Karatkevich. Heurystyki dla metody Thelena obliczania implikantów prostych. In *Materiały IV Krajowej konferencji Metody i systemy komputerowe w badaniach naukowych i projektowaniu inżynierskim, MSK'03*, pages 71–76, Kraków, 2003.

33. J. Bieganowski and A. Karatkevich. Heuristics for Thelen's prime implicant method. *Schedae Informaticae*, 14:125–135, February 2005.

34. K. Bilinski. *Application of Petri Nets in Parallel Controllers Design*. PhD thesis, University of Bristol, 1996.

35. K. Bilinski, M. Adamski, J. Saul, and E. Dagless. Petri-net-based algorithms for parallel-controller synthesis. *IEE Proceedings - Computers and Digital Techniques*, 141(6):405–412, 1994.

36. G. Bruno, A. Castella, G. Macario, and M. P. Pescarmona. Scheduling hard real-time systems using high-level Petri nets. In *Application and Theory of Petri Nets 1992, Lecture Notes in Computer Science*, volume 616, pages 93–112. Springer-Verlag, 1992.

37. P. Buchholz. Hierarchical high level Petri nets for complex system analysis. In Valette, R., editor, *Application and Theory of Petri Nets 1994, Proceedings of 15th International Conference, Zaragoza, Spain, Lecture Notes in Computer Science*, volume 815, pages 119–138. Springer-Verlag, 1994.

38. L. D. Cheremisinova. Minimization of finite-response sequential automata that implement parallel logical control algorithms. *Automatic Control and Computer Sciences*, 22(4):69–74, 1988.

39. L. D. Cheremisinova. Software and hardware implementation of PRALU-algorithms. In *Logical Control*, volume 6, pages 88–105, Minsk, 2001. Institute of Engineering Cybernetics of NANB. (in Russian).

40. L. D. Cheremisinova. *Implementation of Parallel Logical Control Algorithms*. Institute of Engineering Cybernetics of NANB, Minsk, 2002. (in Russian).

41. L. D. Cheremisinova. Optimal state assignment of asynchronous parallel automata. In M. Adamski, A. Karatkevich, and M. Węgrzyn, editors, *Design of Embedded Control Systems*, pages 125–137. Springer-Verlag, New York, 2005.

42. L. D. Cheremisinova and Yu. V. Pottosin. Assignment of partial states of a parallel synchronous automaton. In *Proceedings of the International Conference on Computer-Aided Design of Discrete Devices*, pages 85–88, Minsk-Szczecin, 1995.

43. S. Christensen and N. D. Hansen. Coloured Petri nets extended with place capacities, test arcs and inhibitor arcs. In *Proceedings of the 14th International Conference on Application and Theory of Petri Nets, Lecture Notes in Computer Science*, volume 691, pages 186–205, London, UK, 1993. Springer-Verlag.

44. E. M. Clarke, O. Grumberg, M. Minea, and D. A. Peled. State space reduction using partial order techniques. *STTT*, 2(3):279–287, 1999.

45. E. M. Clarke, O. Grumberg, and D. A. Peled. *Model checking*. MIT Press, Cambridge, USA, 1999.

46. A. Classen. *Modulare Statecharts: Ein formaler Rahmen zur hierarchischen Processspezifikation*. Lehrstuhl für Informatik II, Aachen University of Technology, M.Sc. Thesis, Germany, 1993.

47. J. M. Colom, J. Campos, and M. Silva. On liveness analysis through linear algebraic techniques. In *Proceedings of Design Methods Based on Nets, Esprit Basic Research Action 3148, W.G. 6, Deliverables Covering the Period June 1989 to June 1990*, Paris, 1990.

48. T. H. Cormen, Ch. E. Leiserson, and R. L. Rivest. *Introduction to Algorithms*. MIT Press, 1994.

49. O. Coudert and J. K. Madre. New ideas for solving covering problems. In *Proceedings of the Design Automation Conference, DAC'95*, pages 641–646, 1995.

50. O. Coudert, J. K. Madre, and H. Fraisse. A new viewpoint on two-level logic minimization. In *Proceedings of the Design Automation Conference, DAC'93*, pages 625–630, 1993.

51. R. David and H. Alla. *Petri Nets and Grafcet. Tools for Modelling Discrete Event Systems*. Prentice-Hall, New York, 1992.

52. J. Davis, C. Hylands, J. Janneck, E. A. Lee, et al. Overview of the Ptolemy Project. Technical Memorandum UCB/ERL M01/11, Berkeley, March 2001.

53. N. Deo. *Graph Theory with Applications to Engineering and Computer Science*. Prentice-Hall, New Jersey, 1974.

54. J. Desel. A proof of the rank theorem for extended free choice nets. In *Applications and Theory of Petri Nets 1992, Lecture Notes in Computer Science*, volume 616, pages 134–153. Springer-Verlag, 1992.

55. J. Desel and J. Esparza. *Free Choice Petri Nets*. Cambridge University Press, 1995.

56. D. Drusinsky and D. Harel. Using Statecharts for hardware description and synthesis. *IEEE Transactions on Coputer-Aided Design*, 8(7):798–807, July 1989.

57. J. Esparza. Model checking using net unfoldings. In *Proceedings of TAPSOFT '93, Lecture Notes in Computer Science*, volume 668, pages 613–628. Springer-Verlag, 1993.

58. J. Esparza and S. Römer. An unfolding algorithm for synchronous products of transition systems. In *Proceedings of 10th International Conference on Concurrency Theory, Eindhoven, Lecture Notes in Computer Science*, volume 1664, pages 2–20. Springer-Verlag, August 1999.

59. J. Esparza, S. Römer, and W. Vogler. An improvement of McMillan's unfolding algorithm. In *2nd TACAS, LNCS 1055*, pages 87–106. Springer-Verlag, 1996.

60. J. Esparza and M. Silva. On the analysis and synthesis of free choice systems. In *Advances in Petri Nets 1990, Lecture Notes in Computer Science*, volume 483, pages 243–286, Berlin, Germany, 1991. Springer-Verlag.

61. J. Esparza and M. Silva. A polynomial-time algorithm to decide liveness of bounded free choice nets. *Theor. Computer Science*, 102(1):185–205, 1992.

62. R. Esser. *An Object Oriented Petri Net Approach to Embedded Systems Design*. PhD thesis, Swiss Federal Institut of Technology, Zürich, 1996.

63. P. C. McGeer et al. Espresso-Signature: a new exact minimizer for logic functions. In *Proceedings of the Design Automation Conference, DAC'93*, pages 618–624, 1993.

64. R. K. Brayton et al. VIS: a system for verification and synthesis. In *Proceedings of the Conference on Computer-Aided Verification, CAV'96, Lecture Notes in Computer Science*, volume 1102, pages 332–334. Springer-Verlag, August 1996.

65. J. Ezpeleta, J. M. Colom, and J. Martinez. A Petri net based deadlock prevention policy for flexible manufacturing systems. *IEEE Transactions on Robotics and Automation*, 11(2):173–184, 1995.

66. J. Ezpeleta, J. M. Couvreur, and M. Silva. A new technique for finding a generating family of siphons, traps and st-components. Application to colored Petri Nets. In *Advances in Petri Nets 1993, Lecture Notes in Computer Science*, volume 674, pages 126–147. Springer-Verlag, 1993.

67. W. Fengler, A. Wendt, M. Adamski, and J. L. Monteiro. Petri net based program design and implementation for controller systems. In *Proceedings of 1996 IFAC Triennial World Congress*, volume 1, pages 425–429, San Francisco, USA, 1996.

68. J. Fernandes, M. Adamski, and A. Proença. VHDL generation from hierarchical Petri net specifications of parallel controllers. *IEE Proceedings - Computers and Digital Techniques*, 144(2):127–137, 1997.

69. A. Ferrari. JPVM: network parallel computing in Java. *Concurrency: Practice and Experience*, 10(11–13):985–992, 1998.

70. L. Ferrarini. An incremental approah to logic controller design with Petri nets. *IEEE Transactions on System, Man and Cybernetics*, 22(3):461–474, 1992.

71. D. D. Gajski, F. Vahid, S. Narayan, and J. Gong. *Specification and Design of Embedded Systems*. Prentice-Hall, 1994.

72. A. Geist, A. Beguelin, J. Dongarra, W. Jiang, R. Manchek, and V. Sunderam. *PVM: Parallel Virtual Machine. A Users' Guide and Tutorial for Networked Parallel Computing*. Massachusetts Institute of Technology Press, 1994.

73. A. Gill. *Introduction to the Theory of Finite-state Machines*. McGraw-Hill, New York, 1962.

74. C. Girault and R. Valk. *Petri Nets for Systems Engineering. A Guide to Modeling, Verification, and Application*. Springer-Verlag, Berlin Heidelberg, 2003.

75. V. M. Glushkov. *Synthesis of Digital Automata*. Fizmatgiz, Moscow, 1962. (in Russian).

76. P. Godefroid. Using partial orders to improve automatic verification methods. In *Proceedings of the 2nd International Workshop on Computer Aided Verification CAV '90, Lecture Notes in Computer Science*, volume 531, pages 176–185. Springer-Verlag, 1991.

77. P. Godefroid. *Partial-Order Methods for the Verification of Concurrent Systems: An Approach to the State-Explosion Problem, LNCS*, volume 1032. Springer-Verlag, New York, USA, 1996.

78. P. Godefroid and D. Pirottin. Refining dependencies improves partial-order verification methods (extended abstract). In *Proceedings of the 5th International Conference on Computer Aided Verification, CAV '93, LNCS*, volume 697, pages 438–449, London, UK, 1993. Springer-Verlag.

79. L. Gomes, J. P. Barros, and A. Costa. Structuring mechanisms in Petri net models. In M. Adamski, A. Karatkevich, and M. Węgrzyn, editors, *Design of Embedded Control Systems*, pages 153–166. Springer-Verlag, 2005.

80. V. S. Grigoryev, A. D. Zakrevskij, and V. A. Perchuk. The sequent model of the discrete automaton. In *Vychislitelnaya Tekhnika v Mashinostroenii*, pages 147–153. Institute of Engineering Cybernetics, Minsk, March 1972. (in Russian).

81. A. Gurel, O. C. Pastravanu, F. L. Lewis, and A. Doganalp. Deadlock avoidance using a (min,+) matrix model for flexible manufacturing systems. In *Proceedings of the 4th Workshop on Discrete Event Systems*, Cagliari, Italy, 1998.

82. M. Hack. *Analysis of Production Schemata by Petri Nets*. MIT Project MAC TR-94, 1972. Corrections: Project MAC, Computation Structures Note 17 (1974).

83. D. Harel. Statecharts: a visual formalism for complex systems. *Science of Computer Programming*, 8:231–274, 1987.

84. J. Hartmann, M. Vieira, H. Foster, and A. Ruder. A UML-based approach to system testing. *Innovations in Systems and Software Engineering*, 1:12–24, 2005.

85. M. Heiner. Verification and optimization of control programs by Petri nets without state explosion. In *Proceedings of the 2nd International Workshop on Manufacturing and Petri Nets held at the International Conference on Application and Theory of Petri Nets, ICATPN '97, Toulouse, June 1997*, pages 69–84, 1997.

86. M. Heiner. Petri net based system analysis without state explosion. In *Proceedings of High Performance Computing, Boston, April 1998*, pages 394–403, San Diego, 1998.

87. K. Heljanko, V. Khomenko, and M. Koutny. *Parallelisation of the Petri Net Unfolding Algorithm.* Technical Report CS-TR-733, University of Newcastle upon Tyne, 2001.

88. B. Hnatkowska and Z. Huzar. Transformation of dynamic aspects of UML models into LOTOS behaviour expressions. *Applied Mathematics and Computer Science,* 11(2):537–556, 2001.

89. G. J. Holzmann. *Design and validation of computer protocols.* Prentice-Hall, Upper Saddle River, NJ, USA, 1991.

90. G. J. Holzmann. *The Spin Model Checker: Primer and Reference Manual.* Addison-Wesley, 2003.

91. J. E. Hopcroft and J. D. Ulman. *Introduction to Automata Theory, Languages and Computation.* Addison-Wesley, 1979.

92. Yi. Huang, M. Jeng, X. Xie, and S. Chung. A deadlock prevention policy for flexible manufacturing systems using siphons. In *Proceedings of IEEE International Conference on Robotics and Automation, 2001,* volume 1, pages 541–546, 2001.

93. P. Huber, A. M. Jensen, L. O. Jensen, and K. Jensen. Towards reachability trees for high-level Petri nets. In *Advances in Petri Nets 1984, Lecture Notes in Computer Science,* volume 188, pages 215–233. Springer-Verlag, 1985.

94. Th. Hummel and W. Fengler. Design of embedded control systems using hybrid Petri nets. In *Proceedings of the International Workshop on Discrete - Event System Design,* pages 189–194, Przytok k/Zielonej Góry, Poland, 2001.

95. Th. Hummel and W. Fengler. Design of embedded control systems using hybrid Petri nets and time interval Petri nets. In M. Adamski, A. Karatkevich, and M. Węgrzyn, editors, *Design of Embedded Control Systems,* pages 141–151, New York, 2005. Springer-Verlag.

96. R. Janicki and M. Koutny. *On Some Implementation of Optimal Simulation.* Technical Report 90-07, McMaster University, Hamilton, Ontario, 1990.

97. R. Janicki and M. Koutny. Optimal simulations, nets and reachability graphs. In *Advances in Petri Nets 1991, LNCS,* volume 524, pages 205–226. Springer-Verlag, 1991.

98. R. Janicki and M. Koutny. Using optimal simulations to reduce reachability graphs. In *Proceedings of the 2nd International Conference on Computer-Aided Verification CAV'90, LNCS,* volume 531, pages 166–175. Springer-Verlag, London, 1991.

99. R. Janicki, P. E. Lauer, M. Koutny, and R. Devillers. Concurrent and maximally concurrent evolution of non-sequential systems. *Theoretical Computer Science,* 43:213–238, 1986.

100. K. Jensen. *Coloured Petri Nets. Basic Concept, Analysis Methods and Practical Use. Volume 1: Basic Concepts.* EATCS Monographs on Theoretical Computer Science. Springer-Verlag, Berlin, 1992.

101. K. Jensen and G. Rozenberg, editors. *High-Level Petri Nets: Theory and Application.* Springer-Verlag, Berlin, 1991.

102. A. A. Jerraya and J. Mermet, editors. *System-Level Synthesis.* Kluwer Academic Publishers, 1999.

103. J. Kalinowski. *Wykorzystanie sieci Petriego do projektowania systemów cyfrowych.* PhD thesis, Politechnika Warszawska, 1984.

104. A. Karatkevich. Correctness analysis of α-nets. In *Logical Control,* volume 1, pages 97–106. Institute of Engineering Cybernetics, Academy of Sciences of Belarus, 1996. (in Russian).

105. A. Karatkevich. *Analysis and Optimization of Parallel Logical Control Algorithms*. PhD thesis, Belarusian State University of Informatics and Radioelectronics, Dept. of Computer Science, Minsk, 1997. Manuscript (in Russian).

106. A. Karatkevich. Reachability analysis of the live and safe Petri nets. In *Proceedings of the 5th International Conference the International Conference on Parallel Computing in Electrical Engineering PARELEC'98*, pages 175–178, Białystok, Poland, 1998.

107. A. Karatkevich. Hierarchical decomposition of safe Petri nets. In *Proceedings of the 3rd International Conference on Computer - Aided Design of Discrete Devices*, volume 1, pages 34–39, Minsk, Belarus, 1999.

108. A. Karatkevich. Minimization of transitions number for a parallel automaton. In *Proceedings of the International Conference on Discrete Optimization Methods in Scheduling and Computer-Aided Design*, pages 197–201, Minsk, Belarus, September 2000.

109. A. Karatkevich. Optimal simulation of α-nets. In *Proceedings of the Polish-German Symposium on Science, Research and Education*, pages 217–222, Zielona Góra, Poland, 2000.

110. A. Karatkevich. Dependence between some behaviour properties of α-nets. In *Proceedings of the 4th International Conference on Computer-Aided Design of Discrete Devices*, pages 57–60, Minsk, 2001.

111. A. Karatkevich. On algorithms for decyclisation of oriented graphs. In *Proceedings of the International Workshop on Discrete - Event System Design*, pages 35–40, Przytok k/Zielonej Góry, Poland, 2001.

112. A. Karatkevich. Dynamic reduction of reachability graphs of Petri nets. *Radioelektronika i Informatika*, (no 1):76–82, 2002. (in Russian).

113. A. Karatkevich. Deadlock analysis in Statecharts. In *Proceedings of the Forum on Specification and Design Languages*, pages 414–424, Frankfurt, Germany, September 2003.

114. A. Karatkevich. To behavior analysis of a class of Petri nets. In *Proc. of 27th IFAC/IFIP/IEEE Workshop on Real-Time Programming*, pages 33–38, Lagów, Poland, May 2003. Elsevier, Oxford, UK.

115. A. Karatkevich. Detecting possible deadlock states in automata networks. *Radiotechnika*, nr 138:134–140, 2004. (in Russian).

116. A. Karatkevich. Detection of the unreachable states in FSM networks. In *Proceedings of the 5th International Conference on Computer-Aided Design of Discrete Devices*, pages 47–54, Minsk, 2004.

117. A. Karatkevich. Deadlock detection in discrete concurrent systems. In *Proceedings of the 8th International Conference on experience of designing and application of CAD systems in microelectronics CADSM'2005*, pages 250–253, Lviv-Polyana, Ukraine, 2005.

118. A. Karatkevich. Memory-saving analysis of Petri nets. In M. Adamski, A. Karatkevich, and M. Węgrzyn, editors, *Design of Embedded Control Systems*, pages 65–74. Springer-Verlag, 2005.

119. A. Karatkevich. Properties and analysis of α-nets. *Informatyka Teoretyczna i Stosowana*, 5(8):53–64, 2005.

120. A. Karatkevich. Verification of implementation of parallel automata (testing approach). In *Proceedings of the IEEE East-West Design and Test Workshop EWDTW'05*, pages 66–69, Odessa, Ukraine, 2005.

121. A. Karatkevich. Stubborn set method for interpreted Petri nets. In *Proceedings of the 3rd IFAC Workshop on Discrete-Event System Design, DESDes'06*, pages 227–232, Rydzyna, Poland, 2006.

122. A. Karatkevich. Verification of implementation of parallel automata (symbolic approach). In *Proceedings of the IEEE East-West Design and Test Workshop EWDTW'06*, pages 112–115, Sochi, Russia, 2006.

123. A. Karatkevich and M. Adamski. Deadlock analysis of Petri nets: minimization of memory amount. In *Proceedings of the 3th Electronic Circuits and Systems Conference*, pages 69–72, Bratislava, 2001.

124. A. Karatkevich, M. Adamski, and M. Węgrzyn. Rapid correctness analysis for Sequential Function Chart. In *Proceedings of the 45th International Scienific Colloquium IWK'2000*, pages 679–684, Ilmenau, Germany, October 2000.

125. A. Karatkevich and G. Andrzejewski. Analiza wybranych własności interpretowanej sieci Petriego metodź optymalnej symulacji. In *Materiały Pierwszej Krajowej Konferencji Elektroniki*, volume 2, pages 685–690, Kołobrzeg-Dźwirzyno, Polska, 2002.

126. A. Karatkevich and G. Andrzejewski. Hierarchical decomposition of Petri nets for analysis and design of digital microsystems. *International Journal of Computing*, 5(1):18–25, 2006.

127. A. Karatkevich and G. Andrzejewski. Hierarchical decomposition of Petri nets for digital microsystems design. In *Proceedings of the International Conference on Modern Problems of Radio Engineering, Telecommunications and Computer Science - TCSET'2006*, pages 518–521, Lviv-Slavsk, Ukraine, 2006.

128. A. Karatkevich and T. Gratkowski. Analysis of the operational Petri nets by a distributed system. In *Proceedings of the International Conference on Modern Problems of Radio Engineering, Telecommunications and Computer Science - TCSET'2004*, pages 319–322, Lviv-Slavsk, Ukraine, 2004.

129. A. Karatkevich and M. Węgrzyn. ATPG: Current State and Main Problems. In *Proceedings of the Polish-German Symposium on Science, Research and Education*, pages 205–210, Zielona Góra, Poland, 2000.

130. A. Karatkevich and A. Zakrevskij. Analysis of Petri nets by means of concurent simulation. In *Proceedings of the International Conference on Parallel Computing in Electrical Engineering PARELEC*, pages 87–91, Warsaw, Poland, 2002. IEEE Computer Society.

131. A. Karatkiewicz. Dynamiczna analiza systemów współbieżnych. In *Materiały konferencji naukowej Informatyka - sztuka czy rzemiosło, KNWS '04*, pages 41–46, Zielona Góra, Polska, 2004.

132. A. Karatkiewicz. Wykrywanie blokad w systemach priorytetowych. In *Materiały konferencji naukowej Informatyka - sztuka czy rzemiosło, KNWS '05*, pages 67–72, Zielona Góra, 2005.

133. P. Kemper. Linear time algorithm to find a minimal deadlock in a strongly connected free-choice net. In *Proceedings of the 14th International Conference on Application and Theory of Petri Nets 1993, Chicago, USA, Lecture Notes in Computer Science*, volume 691, pages 319–338. Springer-Verlag, 1993.

134. P. Kemper. *Superposition of Generalized Stochastic Petri Nets and its Impact on Performance Analysis*. PhD thesis, University of Dortmund, Dep. of CS, 1997.

135. P. Kemper and F. Bause. An efficient polynomial-time algorithm to decide liveness and boundedness of free choice nets. In *Proceedings of the 13th International Conference on Application and Theory of Petri Nets, Sheffield, Lecture Notes in Computer Science*, volume 616, pages 263–278, 1992.

136. G. Klas. Hierarchical solution of generalized stochastic petri nets by means of traffic processes. In *Proceedingsof the 13th International Conference on Application and Theory of Petri Nets, Sheffield, Lecture Notes in Computer Science*, volume 616, pages 279–298, 1992.

137. R. König and L. Quäck. *Petri-Netze in der Steuerungstechnik.* Oldenbourg, München, 1988.

138. A. Korotkevich. Analysis of asynchronous systems of simple sequentions. In *Proceedings of the 39th International Scientific Colloquium, IWK'94*, pages 446–452, Ilmenau, Germany, 1994.

139. A. Korotkevich. Checking properties of the asynchronous systems of sequents by using reduction. In *Proceedings of the International Conference on Computer-Aided Design of Discrete Devices (CAD DD'95)*, pages 99–102, Minsk-Szczecin, 1995. Wydawnictwo Uczelniane Politechniki Szczecinskiej.

140. V. Ye. Kotov. *Petri Nets.* Moscow, Nauka, 1984. (in Russian).

141. A. V. Kovalyov. Concurrency relation and the safety problem for Petri nets. In *Proceedings of the 13th International Conference on Application and Theory of Petri Nets 1992, Lecture Notes in Computer Science*, volume 616, pages 299–309. Springer-Verlag, June 1992.

142. A. V. Kovalyov. An $O(|S| \times |T|)$-algorithm to verify if a net is regular. In *Proceedings of the 17th International Conference in Application and Theory of Petri Nets 1996, Lecture Notes in Computer Science*, volume 1091, pages 366–379. Springer-Verlag, June 1996.

143. T. Kozłowski. Petri Net Specification Format (PNSF). Technical report, University of Bristol, 1994.

144. T. Kozłowski, E. Dagless, J. Saul, M. Adamski, and J. Szajna. Parallel controller synthesis using Petri nets. *IEE Proceedings - Computers and Digital Techniques*, 142(4):263–271, 1995.

145. L. M. Kristensen. *State Space Methods for Coloured Petri Nets.* PhD thesis, University of Aarhus, 2000.

146. S. Kumagai, S. Kodama, K. Tsuji, and Y. Nakamura. Preservation of liveness in hierarchical Petri nets. *Electron. Commun. Jpn., Part III, Fundam. Electron. Sci.*, 73(5):8–18, May 1990.

147. O. P. Kuznetsov and G. M. Adelson-Velski. *Discrete Mathematics for Engineer.* Energoatomizdat, Moscow, 1988. (in Russian).

148. G. Łabiak. Modelling Statecharts diagrams by means of Petri nets. In *Proceedings of the 6th International Conference on Advanced Computer Systems*, pages 253–259, Szczecin, October 1999.

149. G. Łabiak. From UML Statecharts to FPGA - the HiCoS approach. In *Proceedings of the Forum on Specification and Design Languages*, pages 354–363, Frankfurt, Germany, 2003.

150. G. Łabiak. Symbolic state exploration of UML Statecharts for hardware description. In M. Adamski, A. Karatkevich, and M. Węgrzyn, editors, *Design of Embedded Control Systems*, pages 73–83. Springer-Verlag, New York, 2005.

151. G. Łabiak. *Wykorzystanie hierarchicznego modelu współbieżnego automatu w projektowaniu sterowników cyfrowych.* PhD thesis, Uniwersytet Zielonogórski, 2005.

152. G. Łabiak and M. Adamski. The method of concurrency matrix generation for statechart-based digital controllers. In *Proceedings of the 11th International Conference on Mixed Design of Integrated Circuits and Systems*, pages 445–450, Szczecin, Poland, 2004.

153. G. Łabiak and A. Karatkevich. Metody specyfikacji, syntezy i weryfikacji hierarchicznych diagramów stanów. *IX Konferencja Naukowa Reprogramowalne Ukşady Cyfrowe, Pomiary Automatyka Kontrola*, (Nr. 7, wyd. spec.):109–111, 2006.

154. G. Łabiak and P. Miczulski. UML statecharts and Petri nets model comparison for system level modelling. *Mezhdunarodny sbornik nauchnykh trudov: Progressivnye technologii i sistemy mashinostroyeniya*, 27:310–314, 2004. Donetsk National Technical University, Ukraine.

155. K. Lautenbach. Linear algebraic calculation of deadlocks and traps. In *Concurrency and Nets, Advances of Petri Nets*. Springer-Verlag, Berlin, 1987.

156. K. Lautenbach and H. A. Ridder. A Completion of the S-invariance Technique by means of Fixed Point Algorithms. Research Report 10-95, Universität Koblenz, 1995.

157. B. Lee and E. A. Lee. Hierarchical Concurrent Finite State Machines in Ptolemy. In *Proceedings of the International Conference on Application of Concurrency to System Design*, pages 34–40, Fukushima, Japan, March 1998.

158. A. Lempel. Minimum feedback arc and vertex sets of a directed graph. *IEEE Trans. Circuit Theory*, CT-13(4):399–403, December 1966.

159. R. W. Lewis. *Programming Industrial Control Systems Using IEC 1131-3*. IEE, London, 1995.

160. M. Makela. *A Reachability Analyzer for Algebraic System Nets*. Research Report A69, Helsinki University of Technology, 2001.

161. H. J. Mathony. *Algorithmische Entwurfstverfahren für Zwei- und Mehrstufige Schaltnetze*. PhD thesis, ITIV, Universität Karlsruhe, 1988.

162. H. J. Mathony. Universal logic design algorithm and its application the synthesis of two-level switching circuits. *Proceedings of the IEE*, 136(3):171–177, 1989.

163. A. Mazurkiewicz. Trace theory. In *Petri Nets: Applications and Relationships to Other Models of Concurrency, Advances in Petri Nets 1986, Part II, Proceedings of an Advanced Course, Bad Honnef, September 1986, Lecture Notes in Computer Science*, volume 255, pages 279–324, New York, USA, 1987. Springer-Verlag.

164. C. McCreary, J. Thompson, H. Gill, T. Smith, and Y. Zhu. *Partitioning and Scheduling Using Graph Decomposition*. Department of Computer Science and Engineering CSE-93-06, Auburn University, 1993.

165. K. L. McMillan. Using unfolding to avoid the state explosion problem in the verification of asynchronous circuits. In *Proceedings of the 4th International Conference on Computer Aided Verification, Lecture Notes in Computer Science*, volume 663, pages 164–177, Montreal, Canada, June 1992. Springer Verlag.

166. D. De Micheli. *Synthesis and Optimization of Digital Circuits*. McGraw-Hill, New York, 1994.

167. P. Miczulski. Calculating state spaces of hierarchical Petri nets using BDD. In *Design of Embedded Control Systems*, pages 85–94. Springer, New York, 2005.

168. P. Miczulski. Weryfikacja i synteza programów dla reprogramowalnych sterowników logicznych z wykorzystaniem funkcji monotonicznych i diagramow BDD. *IX Konferencja Naukowa Reprogramowalne Ukşady Cyfrowe, Pomiary Automatyka Kontrola*, (Nr. 7, wyd. spec.):112–114, 2006.

169. R. Milner. *Communication and Concurrency*. Prentice-Hall, Inc., 1989.

170. M. Minoux and K. Barkaoui. Deadlocks and traps in Petri nets as Horn-satisfiability solutions and some related polynomially solvable problems. *Discrete Mathematics*, 29:195–210, 1990.

171. M. Montalbano. High-speed calculation of the critical paths of large networks. *IBM Systems Journal*, 6(3):163–191, 1967.

172. T. Murata. Petri nets: properties, analysis and applications. *Proceedings of the IEEE*, 77:541–580, April 1989.

173. D. Nazareth, F. Regensburger, and P. Sholz. Mini-Statecharts, A Lean Version of Statecharts. Technical Report TUMŨI9610, Technische Universität München, 1996.

174. R. Nelson. Simplest normal truth functions. *Journal of Symbolic Logic*, 20(2):105–108, 1955.

175. M. Notomi and T. Murata. Hierarchical reachability graph of bounded Petri nets for concurrent-software analysis. *IEEE Trans. Softw. Eng.*, 20(5):325–336, 1994.

176. A. Ohta, K. Tsuji, and T. Hisamura. On liveness of extended partially ordered condition nets. *IEICE Trans. on Fundamentals of Electronics, Communications and Computer Sciences*, E82–A(11):2576–2578, 1999.

177. G. K. Palshikar. An introduction to model checking, 2004. www.embedded.com.

178. Ch. H. Papadimitriou. *Computational Complexity*. Addison Wesley, 1994.

179. J. Pardey, A. Amroun, M. Bolton, and M. Adamski. Parallel controller synthesis for programmable logic devices. *Microprocessors and Microsystems*, 18(8):451–457, 1994.

180. E. Pastor and J. Cortadella. Efficient encoding schemes for symbolic analysis of Petri nets. In *Proceedings of the Conference on Design, Automation and Test in Europe, Paris, France*, pages 790–795, 1998.

181. E. Pastor, O. Roig, J. Cortadella, and R. M. Badia. Petri net analysis using boolean manipulation. In Valette, R., editor, *Application and Theory of Petri Nets 1994, Proceedings 15th International Conference, Zaragoza, Spain, Lecture Notes in Computer Science*, volume 815, pages 416–435. Springer-Verlag, 1994.

182. D. Peled. All from one, one for all: on model checking using representatives. In *Proceedings of the 5th International Conference on Computer Aided Verification, Lecture Notes In Computer Science*, volume 697, pages 409–423. Springer-Verlag, 1993.

183. J. L. Peterson. *Petri net theory and the modeling of systems*. Prentice-Hall, 1981.

184. C. A. Petri. *Kommunikation mit Automaten*. Schriften des IIM Nr. 2, Institut für Instrumentelle Matematik, Bonn, 1962.

185. Yu. V. Pottosin. Generation of parallel automata. In *Methods and Algorithms for Logical Design*, pages 132–142, Minsk, 1995. Institute of Engineering Cybernetics of Academy of Sciences of Belarus. (in Russian).

186. Yu. V. Pottosin. Optimal state assignment of synchronous parallel automata. In M. Adamski, A. Karatkevich, and M. Węgrzyn, editors, *Design of Embedded Control Systems*, pages 111–124, New York, 2005. Springer-Verlag.

187. M. Rauhamaa. *A Comparative Study of Methods for Efficient Reachability Analysis*. Licentiate's thesis, Helsinki University of Technology, Department of Computer Science and Engineering, 1990.

188. W. Reisig. *Petri Nets*, volume 4 of *EATCS Monographs on Theoretical Computer Science*. Springer-Verlag, 1985.

189. R. Rudell and A. Sangiovanni-Vincentelli. Multiple-valued minimization for PLA optimization. *IEEE Transactions on CAD/ICAS*, CAD-6(5):727–750, Sept. 1987.

190. B. Rytsar and V. Minziuk. The set-theoretical modification of boolean functions minimax covering method. In *Proceedings of the International Conference on Modern Problems of Radio Engineering, Telecommunications and Computer Science, TCSET'04*, pages 46–48, Lviv-Slavsko, Ukraine, 2004.

191. V. M. Savi and X. Xie. Liveness and boundedness analysis for petri nets with event graph modules. In *13th International Conference on Application and Theory of Petri Nets 1992, Sheffield, UK, Lecture Notes in Computer Science*, volume 616, pages 328–347. Springer-Verlag, June 1992.

192. K. Schmidt. How to calculate symbolically siphons and traps of algebraic Petri nets. In *Technical Report A39*, pages 1–40. Helsinki University of Technology, 1996.

193. K. Schmidt. Siphons and traps for algebraic Petri nets. In *Proceedings of the Workshop CSP*, pages 157–168, Berlin, Oct 1996.

194. K. Schmidt. Characterizing liveness of Petri nets in terms of siphons. In *Proceedings of the 18th International Conference on Application and Theory of Petri Nets, Lecture Notes in Computer Science*, volume 1248, pages 271–289. Springer-Verlag, June 1997.

195. T. Schober, A. Reinsch, and W. Erhard. Modeling and verification of sequential control paths using Petri nets. In *Proceedings of the International Workshop on Discrete - Event System Design*, pages 41–46, Przytok k/Zielonej Góry, Poland, 2001.

196. T. Schober, A. Reinsch, and W. Erhard. Verification of control paths using Petri nets. In M. Adamski, A. Karatkevich, and M. Węgrzyn, editors, *Design of Embedded Control Systems*, pages 51–62, New York, 2005. Springer-Verlag.

197. J. Sifakis. *Le Controle des Systemes Asynchrones: Concepts, Proprietes, Analyse Statique*. PhD thesis, l'Universite Scientifique et Medicale de Grenoble, 1979.

198. M. Silva. Las redes de Petri: en la Automática y la Informática. *Ed. AC*, 1985.

199. Z. Skowroński. *Translacja specyfikacji funkcjonalnej układów cyfrowych na sieć Petirego dla potrzeb syntezy systemowej*. PhD thesis, Politechnika Szczecińska, Wydział Informatyki, 2000.

200. J. Staunstrup, H. R. Andersen, H. Hulgaard, J. Lind-Nielsen, K. G. Larsen, G. Behrmann, K. Kristoffersen, A. Skou, H. Leerberg, and N. B. Theilgaard. Practical verification of embedded software. *IEEE Computer*, 33(5):68–75, 2000.

201. B. Steinbach and A. D. Zakrevskij. Parallel automaton - basic model, properties and high-level diagnostics. In *Proceedings of the 4th International Workshop on Boolean Problems*, pages 151–158, Freiberg, Germany, 2000.

202. B. Steinbach and A. D. Zakrevskij. Parallel automaton and its hardware implementation. In *Proceedings of the Design and Diagnostics of Electronic Circuits and Systems Workshop*, pages 250–257, Smolenice, 2000.

203. U. Stern and L. D. Dill. Parallelizing the Murphy verifier. In *Proceedings of the International Conference CAV'97, LNCS*, pages 256–267. Springer, 1997.

204. M. E. Szabo, editor. *The Collected Papers of Gerhard Gentzen*. North-Holland, Amsterdam, 1969.

205. B. Taconet and B. Chollot. Grafcet programming on programmable logic controller with logical, ladder and boolean language. *Revue Nouvel Automatisme*, 24(1-2):41–45, 1979.

206. S. Tanimoto, M. Yamauchi, and T. Watanabe. Finding minimal siphons in general Petri nets. *IEICE Trans. on Fundamentals of Electronics, Communications and Computer Sciences*, E79–A(11):1817–1824, 1996.

207. B. Thelen. *Investigation of Algorithms for Computer-Aided Logic Design of Digital Circuits*. PhD thesis, Universität Karlsruhe, 1988. (in German).

208. V. V. Tropashko. Proof of the conjecture of complete reducibility of α-nets. In *Design of Logical Control Systems*, pages 13–21. Institute of Engineering Cybernetics of Academy of Sciences of BSSR, 1986. (in Russian).

209. *OMG Unified Modeling Languag Specification Version 1.5*. OMG, 250 First Avenue, Needham, MA 02494, U.S.A., March 2003.

210. A. Valmari. Eliminating redundant interleavings during concurrent program verification. In *Proceedings of the Conference on Parallel Architectures and Languages Europe, Vol. 2, Lecture Notes in Computer Science*, volume 366, pages 89–103, Berlin, Germany, 1989. Springer-Verlag.

211. A. Valmari. A stubborn attack on state explosion. In *Proceedings of the 2nd International Workshop on Computer Aided Verification, CAV '90, LNCS*, volume 531, pages 156–165, London, UK, 1991. Springer-Verlag.

212. A. Valmari. Stubborn sets for reduced state space generation. In *Advances in Petri Nets 1990, Lecture Notes in Computer Science*, volume 483, pages 491–515. Springer-Verlag, Berlin, Germany, 1991.

213. A. Valmari. On-the-fly verification with stubborn sets. In *Proceedings of the 5th International Conference on Computer Aided Verification, CAV'93, LNCS*, volume 697, pages 397–408, London, UK, 1993. Springer-Verlag.

214. A. Valmari. Compositional analysis with place-bordered subnets. In *Application and Theory of Petri Nets, LNCS 815*, pages 531–547. Springer, 1994.

215. A. Valmari. State of the art report: Stubborn sets. *Petri Net Newsletter*, (46):6–14, April 1994.

216. A. Valmari. Stubborn set methods for process algebras. In *Proceedings of the DIMACS workshop on Partial order methods in verification, POMIV'96*, pages 213–231, New York, USA, 1997.

217. A. Valmari. The state explosion problem. In *Lecture Notes in Computer Science: Lectures on Petri Nets I: Basic Models*, volume 1491, pages 429–528. Springer-Verlag, 1998.

218. V. Varadharajan and K. D. Baker. Directed graph based representation for software system design. *Software Engineering J.*, 2(1):21–28, January 1987.

219. K. Varpaaniemi. On combining the stubborn set method with the sleep set method. In *Proceedings of the 15th International Conference on Application and Theory of Petri Nets, Lecture Notes in Computer Science*, volume 815, pages 548–567. Springer-Verlag, Berlin, Germany, 1994.

220. K. Varpaaniemi. On Computing Symmetries and Stubborn Sets. Technical Report B12, Helsinki University of Technology, Digital Systems Laboratory, 1994.

221. K. Varpaaniemi. *On the Stubborn Set Method in Reduced State Space Generation.* PhD thesis, Helsinki University of Technology, Department of Computer Science and Engineering, 1998.

222. K. Varpaaniemi. Stubborn sets for prioriry nets. In *Proceedings of ISCIS 2004, Lecture Notes in Computer Science*, volume 3280, pages 574–583. Springer-Verlag, Berlin, Germany, 2004.

223. M. von der Beeck. A Comparison of Statecharts Variants. In *LNCS*, volume 860, pages 128–148. Springer-Verlag, 1994.

224. F. Wagner and M. Rakowski. Zastosowanie sieci Petriego do projektowania mikroprogramowych układów sekwencyjnych. *Archiwum Automatyki i Telemechaniki*, 26(3):385–395, 1981.

225. P. Wołański. *Modelowanie układów cyfrowych na poziomie RTL z wykorzystaniem sieci Petriego i podzbioru języka VHDL.* PhD thesis, Politechnika Warszawska, 1998.

226. P. Wolper and P. Godefroid. Partial-order methods for temporal verification. In *Proceedings of the 4th International Conference on Concurrency Theory, CONCUR'93, LNCS*, volume 715, pages 233–246. Springer-Verlag, UK, 1993.

227. A. Węgrzyn and P. Bubacz. XML format for high-level Petri net. In *Proceedings of ACS 2001*, volume 2, pages 269–278, Szczecin, 2001.

228. A. Węgrzyn, A. Karatkevich, and J. Bieganowski. Detection of deadlocks and traps in Petri nets by means of Thelen's prime implicant method. *Applied Mathematics and Computer Science*, 14(1):113–121, 2004.

229. A. Węgrzyn and M. Węgrzyn. Symbolic veification of concurrent logic controllers by means Petri nets. In *Proceedings of CAD DD'99*, volume 1, pages 45–50, Minsk, Belarus, 1999.

230. M. Yamauchi, S. Tanimoto, and T. Watanabe. Finding a minimal siphon containing specified places in a general Petri net. *IEICE Trans. on Fundamentals of Electronics, Communications and Computer Sciences*, E79–A(11):1825–1828, 1996.

231. M. Yamauchi and T. Watanabe. Time complexity analysis of the minimal siphon extraction problem of Petri nets. *IEICE Trans. on Fundamentals of Electronics, Communications and Computer Sciences*, E82–A(11):2558–2565, 1999.

232. S. A. Yuditski, A. A. Tagayevskaya, and G. K. Yefremova. *A Language for Algorithmic Design of Discrete Control Devices*. Preprint of Institute of Control Sciences, Moscow, 1977. (in Russian).

233. S. A. Yuditski, A. A. Tagayevskaya, and G. K. Yefremova. *Design of Discrete Systems of Automatics*. Mashinostroyeniye, Moscow, 1980. (in Russian).

234. V. N. Zakharov. Sequent description of control automata. *Izvestiya AN SSSR*, (2), 1972. (in Russian).

235. A. D. Zakrevskii. Verifying the correctness of parallel logical control algorithms. *Program. Comput. Softw. (USA)*, 13(5):218–221, 1987.

236. A. D. Zakrevskij. A-net - a functional model of a discrete system. *Doklady AN BSSR*, 25(8):714–717, 1981. (in Russian).

237. A. D. Zakrevskij. *Logical Synthesis of Cascade Networks*. Nauka, Moscow, 1981. (in Russian).

238. A. D. Zakrevskij. Reduction method of correctness checking of parallel logical control algorithms. *Doklady AN BSSR*, 27(7):617–619, 1982. (in Russian).

239. A. D. Zakrevskij. Implementation of parallel logical control algorithms on programmable logic arrays. *Automatika i Telemechanika*, (7):116–123, 1983. (in Russian).

240. A. D. Zakrevskij. Parallel automaton. *Doklady AN BSSR*, 28(8):717–719, 1984. (in Russian).

241. A. D. Zakrevskij. To checking liveness of ordinary Petri nets. *Doklady AN BSSR*, 29(11):1006–1009, 1985. (in Russian).

242. A. D. Zakrevskij. Elements of the theory of α-nets. In *Design of Logical Control Systems*, pages 4–12. Institute of Engineering Cybernetics of Academy of Sciences of BSSR, 1986. (in Russian).

243. A. D. Zakrevskij. The analysis of concurrent logic control algorithms. In *Fundamentals in Computation Theory, Lecture Notes in Computer Science*, volume 278, pages 497–500. Springer-Verlag, 1987.

244. A. D. Zakrevskij. Decomposition approach to analysis of parallel logical control algorithms. In *Formal Models of Parallel Computation*. Siberian section of Academy of Sciences of the USSR, Novosibirsk, 1988. (in Russian).

245. A. D. Zakrevskij. To the theory of parallel algorithms of logical control. *Izvestiya AN SSSR, Tekhnicheskaya Kibernetika*, (5):179–191, 1989. (in Russian).

246. A. D. Zakrevskij. Optimization of matrix of partial state assignment for parallel automata. In *Formalization and Automatization of Logical Design*, pages 4–11. Institute of Engineering Cybernetics of Academy of Science of Belarus, 1993. (in Russian).

247. A. D. Zakrevskij. Parallel logical control algorithms: verification and hardware implementation. *Computer Science Journal of Moldova*, 4(1):3–19, 1996.

248. A. D. Zakrevskij. High-level design of logical control devices. In *Proceedings of the 3rd International Conference on Computer-Aided Design of Discrete Devices*, pages 13–18, Minsk, 1999.

249. A. D. Zakrevskij. *Parallel Algorithms of Logical Control.* Second edition, URSS, Moscow, 2003. (in Russian).

250. A. D. Zakrevskij. Using sequents for description of concurrent digital systems behavior. In M. Adamski, A. Karatkevich, and M. Węgrzyn, editors, *Design of Embedded Control Systems*, pages 3–13, New York, 2005. Springer-Verlag.

251. A. D. Zakrevskij, A. G. Karatkevich, and M. A. Adamski. A method of analysis of operational Petri nets. In *Proceedings of the 8th International Conference on Advanced computer systems*, pages 449–460. Kluwer Academic Publishers, Boston, 2002.

252. A. D. Zakrevskiy. Petri nets modeling of logical control algorithm. *Autom. Control & Comput. Sci.*, 20(6):38–45, 1986.

253. M. Zhou and F. DiCesare. *Petri Net Synthesis for Discrete Event Control of Manufacturing Systems.* Kluwer Academic Publishers, Boston, 1993.

Index

Lecture Notes in Control and Information Sciences

Edited by M. Thoma, M. Morari

Further volumes of this series can be found on our homepage:
springer.com

Vol. 356: Karatkevich A.:
Dynamic Analysis of Petri Net-Based Discrete Systems
166 p. 2007 [978-3-540-71464-4]

Vol. 355: Zhang H.; Xie L.:
Control and Estimation of Systems with Input/Output Delays
213 p. 2007 [978-3-540-71118-6]

Vol. 354: Witczak M.:
Modelling and Estimation Strategies for Fault Diagnosis of Non-Linear Systems
215 p. 2007 [978-3-540-71114-8]

Vol. 353: Bonivento C.; Isidori A.; Marconi L.; Rossi C. (Eds.)
Advances in Control Theory and Applications
305 p. 2007 [978-3-540-70700-4]

Vol. 352: Chiasson, J.; Loiseau, J.J. (Eds.)
Applications of Time Delay Systems
358 p. 2007 [978-3-540-49555-0]

Vol. 351: Lin, C.; Wang, Q.-G.; Lee, T.H., He, Y.
LMI Approach to Analysis and Control of Takagi-Sugeno Fuzzy Systems with Time Delay
204 p. 2007 [978-3-540-49552-9]

Vol. 350: Bandyopadhyay, B.; Manjunath, T.C.; Umapathy, M.
Modeling, Control and Implementation of Smart Structures 250 p. 2007 [978-3-540-48393-9]

Vol. 349: Rogers, E.T.A.; Galkowski, K.; Owens, D.H.
Control Systems Theory and Applications for Linear Repetitive Processes 482 p. 2007 [978-3-540-42663-9]

Vol. 347: Assawinchaichote, W.; Nguang, K.S.; Shi P.
Fuzzy Control and Filter Design for Uncertain Fuzzy Systems
188 p. 2006 [978-3-540-37011-6]

Vol. 346: Tarbouriech, S.; Garcia, G.; Glattfelder, A.H. (Eds.)
Advanced Strategies in Control Systems with Input and Output Constraints
480 p. 2006 [978-3-540-37009-3]

Vol. 345: Huang, D.-S.; Li, K.; Irwin, G.W. (Eds.)
Intelligent Computing in Signal Processing and Pattern Recognition
1179 p. 2006 [978-3-540-37257-8]

Vol. 344: Huang, D.-S.; Li, K.; Irwin, G.W. (Eds.)
Intelligent Control and Automation
1121 p. 2006 [978-3-540-37255-4]

Vol. 341: Commault, C.; Marchand, N. (Eds.)
Positive Systems
448 p. 2006 [978-3-540-34771-2]

Vol. 340: Diehl, M.; Mombaur, K. (Eds.)
Fast Motions in Biomechanics and Robotics
500 p. 2006 [978-3-540-36118-3]

Vol. 339: Alamir, M.
Stabilization of Nonlinear Systems Using Receding-horizon Control Schemes
325 p. 2006 [978-1-84628-470-0]

Vol. 338: Tokarzewski, J.
Finite Zeros in Discrete Time Control Systems
325 p. 2006 [978-3-540-33464-4]

Vol. 337: Blom, H.; Lygeros, J. (Eds.)
Stochastic Hybrid Systems
395 p. 2006 [978-3-540-33466-8]

Vol. 336: Pettersen, K.Y.; Gravdahl, J.T.; Nijmeijer, H. (Eds.)
Group Coordination and Cooperative Control
310 p. 2006 [978-3-540-33468-2]

Vol. 335: Kozłowski, K. (Ed.)
Robot Motion and Control
424 p. 2006 [978-1-84628-404-5]

Vol. 334: Edwards, C.; Fossas Colet, E.; Fridman, L. (Eds.)
Advances in Variable Structure and Sliding Mode Control
504 p. 2006 [978-3-540-32800-1]

Vol. 333: Banavar, R.N.; Sankaranarayanan, V.
Switched Finite Time Control of a Class of Underactuated Systems
99 p. 2006 [978-3-540-32799-8]

Vol. 332: Xu, S.; Lam, J.
Robust Control and Filtering of Singular Systems
234 p. 2006 [978-3-540-32797-4]

Vol. 331: Antsaklis, P.J.; Tabuada, P. (Eds.)
Networked Embedded Sensing and Control
367 p. 2006 [978-3-540-32794-3]

Vol. 330: Koumoutsakos, P.; Mezic, I. (Eds.)
Control of Fluid Flow
200 p. 2006 [978-3-540-25140-8]

Vol. 329: Francis, B.A.; Smith, M.C.; Willems, J.C. (Eds.)
Control of Uncertain Systems: Modelling, Approximation, and Design
429 p. 2006 [978-3-540-31754-8]

Vol. 328: Loría, A.; Lamnabhi-Lagarrigue, F.; Panteley, E. (Eds.)
Advanced Topics in Control Systems Theory
305 p. 2006 [978-1-84628-313-0]

Vol. 327: Fournier, J.-D.; Grimm, J.; Leblond, J.; Partington, J.R. (Eds.)
Harmonic Analysis and Rational Approximation
301 p. 2006 [978-3-540-30922-2]

Vol. 326: Wang, H.-S.; Yung, C.-F.; Chang, F.-R.
H_∞ Control for Nonlinear Descriptor Systems
164 p. 2006 [978-1-84628-289-8]

Vol. 325: Amato, F.
Robust Control of Linear Systems Subject to Uncertain
Time-Varying Parameters
180 p. 2006 [978-3-540-23950-5]

Vol. 324: Christofides, P.; El-Farra, N.
Control of Nonlinear and Hybrid Process Systems
446 p. 2005 [978-3-540-28456-7]

Vol. 323: Bandyopadhyay, B.; Janardhanan, S.
Discrete-time Sliding Mode Control
147 p. 2005 [978-3-540-28140-5]

Vol. 322: Meurer, T.; Graichen, K.; Gilles, E.D. (Eds.)
Control and Observer Design for Nonlinear Finite and Infinite Dimensional Systems
422 p. 2005 [978-3-540-27938-9]

Vol. 321: Dayawansa, W.P.; Lindquist, A.; Zhou, Y. (Eds.)
New Directions and Applications in Control Theory
400 p. 2005 [978-3-540-23953-6]

Vol. 320: Steffen, T.
Control Reconfiguration of Dynamical Systems
290 p. 2005 [978-3-540-25730-1]

Vol. 319: Hofbaur, M.W.
Hybrid Estimation of Complex Systems
148 p. 2005 [978-3-540-25727-1]

Vol. 318: Gershon, E.; Shaked, U.; Yaesh, I.
H_∞ Control and Estimation of State-multiplicative Linear Systems
256 p. 2005 [978-1-85233-997-5]

Vol. 317: Ma, C.; Wonham, M.
Nonblocking Supervisory Control of State Tree Structures
208 p. 2005 [978-3-540-25069-2]

Vol. 316: Patel, R.V.; Shadpey, F.
Control of Redundant Robot Manipulators
224 p. 2005 [978-3-540-25071-5]

Vol. 315: Herbordt, W.
Sound Capture for Human/Machine Interfaces: Practical Aspects of Microphone Array Signal Processing
286 p. 2005 [978-3-540-23954-3]

Vol. 314: Gil', M.I.
Explicit Stability Conditions for Continuous Systems
193 p. 2005 [978-3-540-23984-0]

Vol. 313: Li, Z.; Soh, Y.; Wen, C.
Switched and Impulsive Systems
277 p. 2005 [978-3-540-23952-9]

Vol. 312: Henrion, D.; Garulli, A. (Eds.)
Positive Polynomials in Control
313 p. 2005 [978-3-540-23948-2]

Vol. 311: Lamnabhi-Lagarrigue, F.; Loría, A.; Panteley, E. (Eds.)
Advanced Topics in Control Systems Theory
294 p. 2005 [978-1-85233-923-4]

Vol. 310: Janczak, A.
Identification of Nonlinear Systems Using Neural Networks and Polynomial Models
197 p. 2005 [978-3-540-23185-1]

Vol. 309: Kumar, V.; Leonard, N.; Morse, A.S. (Eds.)
Cooperative Control
301 p. 2005 [978-3-540-22861-5]

Vol. 308: Tarbouriech, S.; Abdallah, C.T.; Chiasson, J. (Eds.)
Advances in Communication Control Networks
358 p. 2005 [978-3-540-22819-6]

Vol. 307: Kwon, S.J.; Chung, W.K.
Perturbation Compensator based Robust Tracking Control and State Estimation of Mechanical Systems
158 p. 2004 [978-3-540-22077-0]

Vol. 306: Bien, Z.Z.; Stefanov, D. (Eds.)
Advances in Rehabilitation
472 p. 2004 [978-3-540-21986-6]

Vol. 305: Nebylov, A.
Ensuring Control Accuracy
256 p. 2004 [978-3-540-21876-0]

Vol. 304: Margaris, N.I.
Theory of the Non-linear Analog Phase Locked Loop
303 p. 2004 [978-3-540-21339-0]

Vol. 303: Mahmoud, M.S.
Resilient Control of Uncertain Dynamical Systems
278 p. 2004 [978-3-540-21351-2]

Vol. 302: Filatov, N.M.; Unbehauen, H.
Adaptive Dual Control: Theory and Applications
237 p. 2004 [978-3-540-21373-4]

Made in the USA
Monee, IL
07 July 2026